EVOLUTIONARY STUDIES IN WORLD CROPS

Diversity and change in the Indian subcontinent

EVOLUTIONARY STUDIES IN WORLD CROPS

Diversity and change in the Indian subcontinent

Edited by

SIR JOSEPH HUTCHINSON, C.M.G., Sc.D., F.R.S.
Drapers' Professor of Agriculture Emeritus
University of Cambridge

CAMBRIDGE UNIVERSITY PRESS

CAMBRIDGE UNIVERSITY PRESS
Cambridge, New York, Melbourne, Madrid, Cape Town, Singapore, São Paulo, Delhi

Cambridge University Press
The Edinburgh Building, Cambridge CB2 8RU, UK

Published in the United States of America by Cambridge University Press, New York

www.cambridge.org
Information on this title: www.cambridge.org/9780521117609

First published 1974
This digitally printed version 2009

A catalogue record for this publication is available from the British Library

ISBN 978-0-521-20339-5 hardback
ISBN 978-0-521-11760-9 paperback

Contents

Preface

De Candolle is the father of the study of crop plant evolution. His classic work was the first attempt to set out in order what was known of the botanical history of domesticated plants. It was not until half a century had elapsed that another great botanist, Vavilov, took up the subject and advanced our understanding by his extensive studies of crop plants in their areas of origin, and by the analysis of the distribution of crop plant variability in his classic papers.

Since Vavilov's time there has been no masterly synthesis of our knowledge such as his, but instead a series of individual studies of crop plant species that have broadened our understanding very greatly. In 1962 a series of lectures on Crop Plant Evolution was organised in Cambridge and published as a volume of essays. They contribute to our knowledge of the range of plants on which there are specialists in or near Cambridge. They do not give a broad picture, but this is not the time to paint on a large canvas. In the present state of knowledge surveys of limited areas of the field are more appropriate.

When I was in India from 1933 to 1937, I was impressed above all else with the wealth of material of genetic interest in the long-established crop plant populations of Indian agriculture. When I had the opportunity to return to India, I thought immediately of this wealth, and of the new knowledge that has been built up and its significance for the study of crop plant evolution. I hoped that we might get together the experts in this field and make a collaborative study of the evolution that has gone on in this subcontinent. I found that the time was even more opportune than I had supposed. Not only is there a great wealth of knowledge of the subject, but this is a critical period. For perhaps 4000 years crop plants have been subjected to a farming environment that only changed slowly. In the last few years the environment to which they must adapt has changed very greatly and is in process of changing even more. So we are in a most favourable position to review the results of past selective forces, and to survey the future and the selective forces it will probably bring to bear.

This was the reason for holding a symposium on Crop Plant Evolution at the I.A.R.I. in New Delhi on 10–13 March 1970. It began with a consideration of the history of agriculture. One of the great advances in our knowledge has been the dating by archaeologists of the beginnings of agriculture and of some of the major events in its history. It is possible now, as never before, to set crop plant evolution against the time scale of agricultural development. So we can determine the rate of evolution with some precision, and we can safeguard outselves against historical reconstructions that would have taken more time than was actually available.

There followed consideration of a range of the crops that have evolved in India. They are of diverse origins. First are those, like wheat, that were among the earliest domesticates in west Asia and were brought to India, probably by the first farming immigrants. Then there are the indigenous Indian crops, and the crops of African origin, domesticated in these two continents from the wild, often with the descendents of their wild ancestors growing alongside them. There are some crops that were brought from southeast Asia. Finally, there are the comparatively recent introductions from the Americas. All save those from the Americas are ancient components of Indian agriculture. Indeed, they are so long-established that only in recent years have their origins been worked out. Even now, there is considerable uncertainty about the provenance of some of the crops of the Indo-African group, and it seems preferable for the present not to attempt a definitive classification into an Indian and an African group.

It is doubtful whether any country, even so large a country as India, has so diverse a repertoire of crops. They have all changed and developed in response to the selective forces of Indian agriculture. This was the material for the symposium. Not all Indian crops were discussed. In some, the greater part of the evolutionary changes following domestication have taken place outside India. Some have recently been the subject of authoritative studies published elsewhere. The accounts here published are intended to complement other studies in crop plant evolution, and to set out the impact of the great agrarian cultures of India on the wide range of crop plants on which they depended.

St. John's College   J.B. Hutchinson
Cambridge 1973

1
The beginnings of Agriculture

Palaeobotanical evidence in India

VISHNU-MITTRE

Birbal Sahni Institute of Palaeobotany, Lucknow

Introduction

The earliest evidence of agriculture so far discovered in India may be dated to 2300 B.C., as is apparent from the radiocarbon dates of the Neolithic sites in the Kashmir Valley, Mysore and Andhra Pradesh. The inclusion of the extreme radiocarbon dates from Kashmir and peninsular India (TF-573, 2905 ± 100 B.P. and TF-748, 4410 ± 105 B.P.) reveals that the Neolithic period in India commenced about 2500 B.C. and continued to 1000 B.C. (Agrawal and Kusumgar, 1968b; Agrawal, Gupta and Kusumgar, 1969). Surprisingly, within this very chronological span, particularly between 2300 and 1750 B.C., Harappan culture flourished in Sind, Punjab, Rajasthan and Saurashtra. This had evolved from a pre-Harappan culture which existed slightly earlier in Sind and Rajasthan. With the Harappan culture emerged the use of copper in addition to stone – the Chalcolithic period. An elaborate town planning and drainage system characterised the Harappan Chalcolithic culture and distinguished it from the post-Harappan Chalcolithic cultures which spread between 1800–1400 B.C. in Rajasthan, Saurashtra and Madhya Pradesh after the decline of the Harappans about 1750 B.C. However scanty the evidence is, it does suggest that the post-Harappan Chalcolithic cultures emerged as a synthesis of the declining culture with new local or exotic cultures.

From Sind, Punjab, Rajasthan, Saurashtra and Madhya Pradesh the radiation of the Chalcolithic cultures proceeded southwards through Deccan, eastwards to the hills of central India and towards the east coast from Andhra to Bengal, and northeastwards into the Ganges valley. This expansion continued until after the emergence of the Iron Age in about 1000 B.C. The Iron Age in India ended about 450 B.C. Thereafter began the Early Historical period to which much of our ancient literature belongs. The advent of the Aryans into India is now believed to date from the late Chalcolithic and early Iron Age (Vishnu-Mittre, 1970).

The above simple outline of the origin and progressive development of Indian cultures immediately brings out the coexistence of primitive and highly advanced cultures both in time and space. The position at the present time in this regard is much as in the past. People in the Jeypore tract of Orissa and other parts of India are still at the stage of food gathering coexisting with highly developed mechanised farming involving improved varieties of cultivars. Thus, the problem of the history of cultivated plants in India is far more complex than elsewhere. It is further complicated by an insufficient knowledge of the material evidence of the emergence and progressive evolution of Indian cultures from pre-existing ones. It appears that the various cultures in India have been isolated entities. Not only were their contacts within India scarce; but there is also little information on the exotic influences on indigenous cultures in the country. Perhaps the assessment has hitherto been too poor. Nevertheless stray material evidence has been evaluated to reveal the contacts of the ancient Indians with the outside world; for instance, of the Neolithic Kashmiris with north China and Mongolia; of Harappans with Baluchistan, Afghanistan, Iran, Caucasus and Mesopotamia; of post-Harappans with Caucasus, Iran and Anatolia; and of Iron Age Indians with Caucasus and Iran. Likewise the Neolithic south Indians are believed to be of caucasoid or of Mediterranean origin with their craft similar to that of northeastern Iran and their stone axe industry reminiscent of the Kashmir Neolithic rather than of Iran. Chinese, Burmese and southeast Asian influences have been inferred for the Neolithic Assamese. It must be pointed out that even a faint indication of any direct or indirect contact with Africa is lacking,[*] though crops of African origin were established in India at an early date. The remarkable discovery of a brick dockyard connected by a channel to the gulf of Cambay at Lothal in Gujarat does, however, suggest a trading post of ancient India, but with what countries marine trading contacts had been established remains to be evaluated. This dockyard had fallen into disuse by about 1800 B.C.

The time and theatre of the origin and progressive development of Indian cultures has much to do with the beginnings of domestication and progressive development of the cultivated plants in India. The time of the earliest adoption of crops in India, as estimated at the moment, is less than 5000 years ago and only about 2400 years B.C. The earliest domestication in western Asia is about 6000 years earlier. It is possible to imagine that some cultivars which were domesticated and evolved in western Asia may have reached India in an advanced form at the beginnings of Indian agriculture. On the other hand, those characteristic of southeast Asia and domesticated there must have reached India much later in time. There are yet others for which India is looked upon as one of the centres of origin and these might have been domesticated here. A continuous history of domestication and diffusion within the country or from outside can be built up only if the evolution of cultures presents an integrated story. This is lamentably lacking in India.

A far more serious limitation than this is the one presented by the materials unearthed from archaeological sites. Information on the food plants of the ancient Indians is to be found in the form of impressions on pottery, and so far rice and sorghum have been found mixed with potter's clay. There are limitations in the study of this kind of material; all the morphological details are not preserved and further it is not possible to measure all the necessary dimensions. The other category of materials comprises carbonised spikelets or kernels, seeds or fruits preserved often in such a high state of carbonisation that important morphological details are lost. Ears, paleae, fragments of spikelet bases, etc. are absent or rarely and poorly preserved. Thus, shape and size of the grain are of limited value for the botanical identification of the material. This can, however, be substantiated through statistical

---

[*] Nagaraja Rao (1971) draws attention to the similarity of pottery head-rests associated with the burials in some Indian archaeological sites such as at Hallur, T. Narsipur with those from Egypt dated to 1400 B.C. He also refers to the use of wooden head-rests in some of the southeast Asian countries, and by the Makabanga and Masozona tribes of southern Rhodesia, and by the Balubas of Zaire.

evaluation of the dimensions. Prior to their comparison with the living counter-parts, changes in dimensions owing to carbonisation must be taken into account. No experiments have yet been conducted in India in this regard.

The time-scale given here is based upon radiocarbon dates (half life cycle: 5730 ± 40 years). In a few instances only have carbonised food grains been radiocarbon dated and most of the dates are based on charcoals. The post-growth factor has not been evaluated and this may be one of the causes for quite a few dates which are looked upon as erratic. These sometimes pose a vexing problem. For instance, rice was introduced at Navdatoli, Madhya Pradesh in Period II dated to 2299–1650 B.C., whereas Period I preceeding it is dated to 1660–1500 B.C. In some cases exact levels for the occurrence of food remains have not been dated for want of sufficient organic material for radiocarbon assay. In such instances the date given is approxi-mate, either calculated purely archaeologically or in relation to preceeding or succeeding ^{14}C dates. The range in dates either signifies records of food plants throughout the time range or it is presumed, as at Mohenjodaro, Harappa, etc., that during the stated time range plant economy remained the same.

Within the framework of the origin and evolution of ancient cultures in India, and the limitations discussed above, this paper describes evidence hitherto unearthed from archaeological sites. Information from pollen-analytical investigations is also given but the event of the earliest landnam phase (occupation phase) remains to be radiocarbon dated. The information hitherto given is all based upon palaeobotanical or palynological investigations. Literary records dealt with earlier (Vishnu-Mittre, 1968c, 1970) are not touched upon here since they all pertain to the early historical period.

Commencement of farming in India

Farming necessitates disturbance of natural vegetation. The events of the earliest disturbance of natural vegetation by Prehistoric Man to clear land for farming can be depicted through pollen analysis of lake and swamp deposits. The shifts in natural vegetation exhibited by pollen diagrams may be conditioned by changes in soil or climate or by the biotic factor. Short-term disturbances in natural vegetation seen in pollen diagrams, such as the decline of forest, rise in grass pollen, appearance of pollen of weeds and also of cereals, and finally the recovery of the damaged forest, are usually interpreted as indicating the influence of man upon vegetation. This disturbance may be accompanied by the presence of charcoal fragments in the corresponding levels of the lake sediments, suggesting clearance of natural vegetation by fire. Such events have been noted in the pollen diagrams constructed from some parts of the country.

In stage b of the pollen diagram constructed from Haigam Lake in the Kashmir Valley (Vishnu-Mittre, 1966b; Vishnu-Mittre and Sharma, 1966) the decline of pine forest, the appearance of *Plantago lanceolata,* the rise of Chenopodiaceae and Compositae shrubs, and subsequently of light-demanding trees, elm and ash, and finally the recovery of pine forest noted at about 5.75 m below the surface of the lake, are suggestive of the clearance of pine forest, the practice of farming and the abandonment of the site by the nomadic neolithic Kashmiris. A similar earlier disturbance at the transition of stages a and b might also be due to change in climate.

At this transition substantial elm decline is noted and the deciduous oak woods are replaced by pine woods. In a similar sequence at the lower part of stage *c* the oak woods are affected. These early evidences of farming are undoubtedly of Neolithic age, dated in the valley as 2375 ± 120 to 1725 ± 95 B.C. (The Neolithic in the valley extends to 1500 B.C., see Agrawal and Kusumgar, 1966.) From the extrapolation of the earlier date, and the rate of sedimentation, maximum farming accompanied by large scale deforestation has been estimated to have occurred during A.D. 700 to 1500. It has, however, not been possible to tell the kind of crops grown but from the nature of the pollen of weeds, wheat or barley can be inferred. A spikelet of rice and pollen of maize appear after A.D. 1500 (Vishnu-Mittre and Sharma, 1966; Vishnu-Mittre and Gupta, 1966). Seeds of *Lithospermum arvense, Medicago denticulata, Medicago* spp. *Lotus corniculatus* and *Ipomoea* sp. discovered from Burzahom, the Neolithic site in the Kashmir Valley (Vishnu-Mittre, 1968*a*), further confirm the inference drawn above of the cultivation of wheat or barley.

Rice is predominantly cultivated in the valley today and in some parts buckwheat is also grown. Pollen evidence has yet to date the shift to rice cultivation and the introduction of buckwheat in the valley.

A pollen diagram from the Naukutchiya Tal near Nainital in Kumaon Himalaya has also furnished more or less similar information (Vishnu-Mittre, Gupta and Robert, 1967). Here the destruction of the pine forest by fire, as indicated by the presence of charcoal in the sediments, precedes the first appearance of cereal pollen. Events in both the diagrams remain to be radiocarbon dated.

The appearance of cereal-type pollen preceded by a short-term shift in pollen curves (namely the decline in pollen curves of grasses, shrubs and trees and rise in pollen curves of *Artemisia,* Umbellifereae, Urticaceae and Compositeae) towards the top of stage *b* in the pollen diagram from Kakathope, Ootacamund, in Madras State at a depth between 200 and 230 cm and dated by radiocarbon to about 23 000 years B.P. (Vishnu-Mittre and Gupta, 1971), immediately shook our confidence in the inference that may be drawn of former cultivation from a short term disturbance as seen in the pollen diagrams. There is a corresponding rise in the pollen curves of the aquatics and plants of the moist habitat, suggesting the influence of factors other than Man. Furthermore, the radiocarbon date is that of the last pluvial. A similar misleading short-term disturbance in vegetation, appearing very much like the Neolithic forest clearance, is known earlier from the Hoxnian Interglacial in England (West, 1956).

Chanda and Mukherjee (1969) have recently reported the presence of a large quantity of grass pollen of a cultivated variety associated with pollen of *Plantago,* etc. from the pollen analysis of superficial peat deposits of Bengal dated about 5000 B.P. If this evidence is established, the earliest record of cultivation in Bengal is dated about 3000 B.C., but the evidence here is not beyond doubt. *Plantago* is a weed of wheat or barley fields. The inference that these two cereals were cultivated in Bengal in Neolithic times is contrary to archaeobotanical evidence. Furthermore, these cereals, being self-pollinated, do not shed sufficient pollen to be carried away and deposited in swamps in large quantity. The authors have not accounted for the occurrence of *Oryza coarctata,* or other wild rices in Bengal marshes of which pollen in large quantity could be found. The maximum size of unacetolysed pollen of *O. coarctata, O. perennis,* wild *O. sativa* and *O. sativa* var. *spontanea* easily reaches 48–53 μm (Misro and Rath, 1961) and acetolysis will bring a further increase in size, thus misleading a pollen analyst into thinking that he has encountered cereal pollen. Thus, agricultural practice as inferred by the authors at 3000 B.C. in Bengal remains unestablished.

In pollen diagrams constructed from Sambhar Lake, Rajasthan, the first cereal type pollen grain of the size range 40–50 μm encountered in stage *b* is dated to 6000 B.C. (Singh, 1967, 1970, 1971), but it is not accompanied by any recognisable disturbance, even in ground vegetation. The cereal-like pollen here must belong to a species of wild grass of which pollen is as large as that of some cultivated cereals. However Singh (1970, 1971) makes a strong case for the earliest farming activity here on the evidence of numerous microscopic fragments of charcoals, in spite of the fact that the vegetation as seen in the pollen diagram remains unaffected by this burning episode. Were the charcoal fragments derived by wind action from the hearth of prehistoric man living beside the shores of the lake? Likewise the 58 per cent cereal pollen from the terrestrial sediments of Kalibangan, an Harappan site in Rajasthan (Singh, 1971), cannot belong to cereals since, owing to prevalence of autogamy among the cereals except *Secale cereale,* it is not possible to recover a high percentage of cereal pollen in soil and lake sediments (Iversen, 1941; Faegri and Iversen, 1964). The archaeobotanical material from this site and others referred to elsewhere in the text does not support the cultivation of this cereal in ancient India. Towards the top of the same diagram Singh (1967, 1970, 1971) deduces intensive clearance and cultivation at a depth dated by radiocarbon to 2700 B.C., corresponding to the Harappan civilisation. Surprisingly at this depth the pollen curve of *Artemisia* attains maximum values and pollen frequencies of Gramineae, though fluctuating, are consistently high. Trees also attain high values opposite the level at which charcoal fragments are found. Thus this pollen diagram hardly reflects any evidence of cultivation as interpreted by Singh. The pollen evidence here is perhaps suggestive of pastoral rather than arable activity, and the avoidance of unpalatable plants of *Artemisia* by the grazing animals may be responsible for the high values here of its pollen grains. The fluctuating high values of grass pollen is in accord with this explanation.

Investigation of the pollen morphology of Indian cereals and of as many as fifty-one wild grasses including the progenitors of cereals (Vishnu-Mittre, 1971*b*) has revealed that on size statistics alone, which is one of the important criteria on which to distinguish the pollen of cereals from that of wild grasses, it is not possible, at least in India, to distinguish the two. Pollen of *Coix lacryma-jobi* is almost as large as that of maize, the largest known in the grasses. The rest of the cereals, with pollen within the range 24–70 μm, find counterparts in the wild grasses of India. Thus, large graminaceous pollen alone is of no use in determining the existence of farming from the Indian pollen diagrams unless accompanied by evidence of disturbance of natural vegetation, appearance of pollen of weeds and the recovery of the forest.

Results of phase-contrast and electron microscopy of pollen of some cereals have been published (Erdtman, 1956; Grohne, 1957; Erdtman, Praglowski and Radwan, 1959; Rowley, 1960), but unless similar results of pollen of many cultivated and wild grasses are published the distinction between cereal and non-cereal pollen will remain uncertain. In regions like that of Europe the strength of Gramineae is much smaller than in India and pollen morphology has established there that size criteria distinguish cereal pollen (but not the pollen of some millets) from that of wild grasses. But these criteria do not hold in India, as the palynological examination of a few out of several hundred species of Gramineae has revealed (Vishnu-Mittre, 1971b). Thus, the detection of the earliest commencement of farming in India through pollen-analytical investigation seems to have been successful only in the Kashmir Valley and Kumaon Himalaya, subject of course to dating by radiocarbon.

Wheat

The Neolithic wheat referred to *Triticum sphaerococcum* has recently been discovered from Chirand, dist. Sarna, Bihar in the Gangetic Plain (Vishnu-Mittre, 1971c, 1972). The depth at which wheat grains are found is much below the depth dated by radiocarbon to 1755 B.C. (TF-1032, 3705 ± 155, Agrawal, 1971). Narain (1970) believes that the Chirand finds may be as old as *c.* 2500 B.C. From the rate of deposition of sediments from the overlying radiocarbon-dated levels, my estimate of the age of the Chirand wheat and other associated food grains is around 3500 B.C. (Vishnu-Mittre, 1971c, 1972). The indirect inference of wheat or barley cultivation from the seeds of weeds of *Lithospermum arvense, Medicago denticulata, Medicago* spp. *Lotus corniculatus* and *Ipomoea* sp. from Burzahom, the Neolithic site in the Kashmir valley dated by radiocarbon to about 2300–1500 B.C. (Vishnu-Mittre, 1968a, b, c), has been referred to earlier.

During the Harappan Period dated by radiocarbon about 2300–1750 B.C. (Agrawal, 1964) wheat is known from Mohenjo-daro and Chanhu-daro in the Sind valley, and from Harappa in Punjab (Marshall, 1931; Luthra, 1936; Vats, 1940; Vishnu-Mittre, 1968b, c). Carbonised wheat grains from the late levels of Mohenjo-daro have been radiocarbon dated to 1755 B.C. (Agrawal, Kusumgar, Lal and Sarna, 1964). The charred grains reported as wheat from Kalibangan, Rajasthan and dated 2090–2075 B.C. (Agrawal and Kusumgar, 1968a, b) have been found on examination to be exclusively barley. Although so far discovered from the close of Harappan levels, we may presume that wheat was one of the staple articles of diet during the Harappan Period.

Recently wheat dated to 1200 B.C. and belonging to post-Harappan times has been found at Atranji Khera, U.P. in the Gangetic Plain (Buth and Chowdhury, 1971).

From the post-Harappan Chalcolithic sites in Madhya Pradesh carbonised wheat dated from 1660 to 1440 B.C. has been discovered at Navdatoli-Maheshwar (Vishnu-Mittre, 1962) and from Kayatha it is known from 1380 B.C. (Agrawal and Kusumgar, 1968b, 1969b). A little later it is known from Sonegaon and Inamgaon in Maharashtra about 1340 to 1290 B.C. (Agrawal and Kusumgar, 1969a, b). Thereafter towards the beginning of the Christian era, about 155 B.C. to A.D. 100, it is known from Ter, dist. Osmanabad, Maharashtra (Vishnu-Mittre, Prakash and Awasthi, 1972). Dating and geographic distribution are summarised in Map 1.

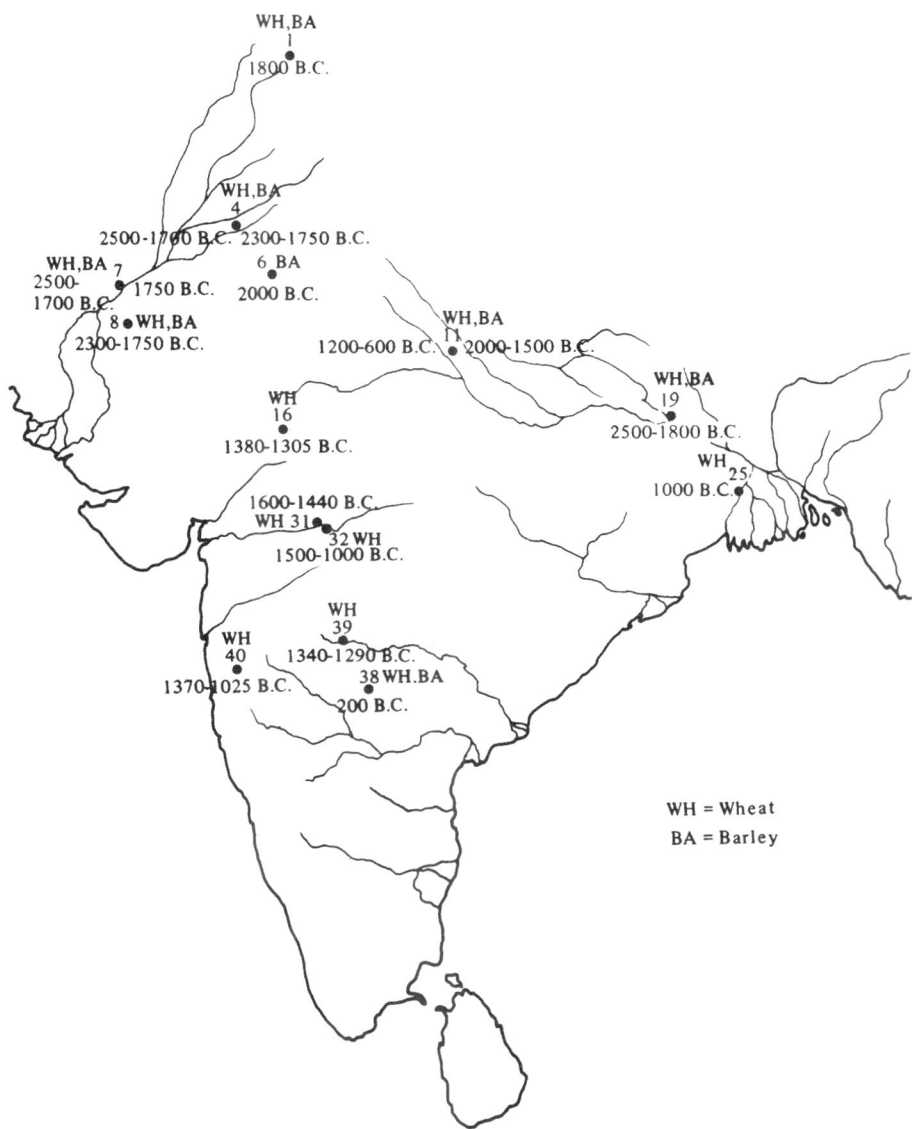

WH,BA
1
1800 B.C.

WH,BA
4
2500-1700 B.C. 2300-1750 B.C.

WH,BA 7
2500-
1700 B.C.
1750 B.C.

6 BA
2000 B.C.

8 WH,BA
2300-1750 B.C.

WH,BA
1200-600 B.C. 2000-1500 B.C.

WH,BA
19
2500-1800 B.C.

WH
16
1380-1305 B.C.

WH 25
1000 B.C.

1600-1440 B.C.
WH 31
32 WH
1500-1000 B.C.

WH
39
1340-1290 B.C.
38 WH,BA
200 B.C.

WH
40
1370-1025 B.C.

WH = Wheat
BA = Barley

Map 1. Geographical distribution of records together with dates of wheat and barley in the Indian subcontinent.

Key to sites plotted on Maps 1–3

1	Burzahom	8	Chanhu-daro	15	Garh Kalika, Ujjain	22	Oriyup
2	Kangra	9	Khokhra Kot	16	Kayatha	23	Sonpur
3	Rupar	10	Hastinapur	17	Kausambhi	24	Mahesdal
4	Harappa	11	Atranji Khera	18	Rajghat	25	Pandu Rajar
5	Rangmahal	12	Noh	19	Chirand		Dhibi
6	Kalibangan	13	Ahar	20	Rajgir	26	Singhbhum
7	Mohenjo-daro	14	Nagda	21	Pataliputra	27	Baidipur

28	Ambri	33	Bhatkuli (Amraoti)	37	Nevasa	41	Kolhapur
29	Lothal	34	Kaundinyapur	38	Ter	42	Hallur
30	Rangpur	35	Pauni	39	Sonegaon	43	Kunnatur
31	Maheshwar	36	Paunar	40	Inamgaon	44	Periyapuram
32	Navdatoli						

Key to symbols for crops on Maps 1–3 and Fig. 1.

BA	Barley, *Hordeum*	MU	Mung, *Phaseolus*
BM	Bajra, *Pennisetum*	PE	Peas, *Pisum*
CA	Castor, *Ricinus*	PH	*Phyllanthus*
CO	Cotton, *Gossypium*	PS	*Paspalum scrobiculatum*
DP	Date palm, *Phoenix*	RA	Ragi, *Eleusine*
HG	Horse gram, *Dolicos*	RI	Rice, *Oryza*
JO	Jowar, *Sorghum*	SE	Sesame, *Sesamum*
LE	Lentils, *Lens*	WH	Wheat, *Triticum*
LI	Linseed, *Linum*	WD	Weeds
MA	Maize, *Zea*	ZI	*Zizyphus*
MS	Melon seeds, *Cucumis*		

Carbonised wheat from Mohenjo-daro was referred to *T. vulgare* and *T. compactum* by Luthra (1936) and Harappa wheat to *T. compactum* or *sphaerococcum* (shot wheat) by Percival (Marshall, 1931). The Navdatoli-Maheshwar wheat was referred to *T. vulgare, compactum* type by Vishnu-Mittre (1962). Re-examination of the charred wheat grains, supported by statistical evaluation of the various dimensions and their comparisons with those of modern grains of *T. sphaerococcum, T. aestivum,* and *T. compactum* (Table 1) shows that the charred wheat grains from Mohenjo-daro, Navdatoli-Maheshwar and Ter compare with those of *T. sphaerococcum,* rather than with the other species. Differences noted may be due to the effect of carbonisation on dimensions. The Navdatoli and Mohenjo-daro grains are comparatively larger and broader than those of Ter, Maharashtra. That is the only change noted within a span of time spread over 2000 years. The single carbonised wheat grain from the Neolithic site, Chirand, Bihar measures 3.75 mm long, 2.75 mm broad and 1.75 mm thick. In its lesser L/B (1.36) and L/T (2.14) and higher B/L (0.73) and T/L (0.46) indices it compares, besides being short, deeply grooved and with the domed dorsal surface, with the modern grain of *T. sphaerococcum* (Vishnu-Mittre, 1972).

From the above it seems that the ancient wheat in India was *T. sphaerococcum,* a hexaploid, which is even today basically adapted to northwestern India. Nothing is at present known about the process of its domestication. Cross-breeding of a wild form of emmer (*T. dicoccum*) and a wild species of *Aegilops* is believed to have given rise to *T. sphaerococcum*. Wild emmer is absent from northwest India today. *Aegilops tauschi* is reported to occur in Kashmir (Bor, 1960).

The distribution of finds of wheat recorded in Map 1 shows that most records come from the western part of the subcontinent, namely from Sind, Punjab, U.P., Madhya Pradesh and Maharashtra. It is amazing that, so far, no records are available from Rajasthan and Saurashtra, where remains of the Harappan culture have also been found. Two eastern records are from Chirand and Pandurajah Dhibi (Director of Archaeology, Calcutta, personal communication).

Table 1 *Dimensions of wheat grains (modern and carbonised)*

	Length L (mm)			Breadth B (mm)			Thickness T (mm)			L/B	L/T	B/L	T/L	T/B
	Av.	Min.	Max.	Av.	Min.	Max.	Av.	Min.	Max.					
T. aestivum	6.22	4.50	7.00	2.32	1.50	3.50	2.02	1.00	2.50	2.68	3.07	0.37	0.32	0.87
T. compactum	5.80	5.00	6.25	2.37	1.88	3.00	1.88	1.50	2.50	2.44	3.07	0.40	0.32	0.79
T. sphaerococcum	4.40	3.25	5.00	2.50	2.00	3.00	2.70	2.50	3.00	1.76	1.62	0.56	0.61	1.09
TER 4/137	4.94	4.00	5.50	2.59	2.25	3.00	2.03	1.50	2.50	1.90	2.21	0.52	0.41	0.78
TER 1396	4.85	4.25	5.50	2.53	2.00	3.00	2.56	2.00	3.00	1.91	2.13	0.52	0.46	0.89
TER 1314	4.75	4.00	6.00	2.42	1.75	3.00	1.66	1.00	2.50	1.96	2.26	0.50	0.35	0.68
TER 5/157	4.74	4.00	5.50	2.61	2.00	3.00	1.97	1.50	3.00	1.81	2.40	0.54	0.41	0.75
TER 433	4.69	4.00	5.50	2.60	2.00	3.50	2.25	1.50	3.00	1.80	2.08	0.55	0.47	0.86
TER 533	4.66	4.00	5.50	2.76	2.00	3.00	2.45	1.50	3.00	1.69	1.90	0.55	0.42	0.89
TER 964	4.65	4.00	5.50	2.57	2.00	3.00	1.95	1.50	2.50	1.80	2.38	0.58	0.50	0.75
TER 5/134	4.62	4.00	5.50	2.70	2.00	3.00	2.34	1.75	3.00	1.71	1.97	0.58	0.50	0.86
TER 1137	4.52	3.50	5.50	2.91	2.25	3.50	2.45	2.00	3.00	1.55	1.84	0.64	0.54	0.84
TER 5/206	4.49	3.50	5.50	2.65	2.00	3.00	2.02	1.00	2.50	1.69	2.22	0.59	0.45	0.76
TER 980	3.60	4.00	6.00	2.50	2.00	3.00	1.85	1.50	2.50	1.44	1.90	0.69	0.51	0.74
Average for Ter material										1.75	2.12	0.57	0.46	0.80
Navdatoli-Maheshwar	6.30	6.10	6.80	3.90	3.70	4.00	3.10	2.50	3.90	1.61	2.03	0.61	0.49	0.79
	5.50	5.10	6.20	3.60	3.30	4.00	2.60	2.00	3.10	1.52	2.10	0.65	0.46	0.72
	4.60	3.90	5.30	2.90	2.30	3.60	2.70	2.10	3.30	1.59	1.70	0.63	0.59	0.93
Mohenjo-daro	5.35	3.25	6.00	2.25	1.25	3.25	2.00	1.25	2.75	1.91	2.39	0.54	0.46	0.81

Rice

A mass of information on the antiquity of rice in India is derived from the imprints of rice spikelets and kernels on potsherds. Either there was an ancient practice of mixing spikelets, grains and chaff of rice with clay as a binding material, prior to turning it into pottery, or else the marshes, which were the source of clay, abounded in wild rices and inadvertently contributed spikelets into the clay. In the Jeypore tract of Orissa, where wild rice grows in abundance, the surface of the marsh is densely strewn with spikelets of rice.* Examination of a handful of clay from marshes bearing *Oryza perennis* at Tulsipur and those bearing *O. sativa* var. *spontanea* from close to Balianta Road in Cuttack revealed 7—10 spikelets in each but the marsh mud contained, besides these, molluscs and remains of plants and animals growing in them. Of these there is no evidence in the potsherds. Nothing but spikelets or chaff are found. This indicates that the practice was intentional, but not widespread since pottery at several sites is devoid of any organic matter. Sometimes very coarse and gritty material without any binding substance has been found turned into pottery as at Bagor, a Neolithic site in Rajasthan. Spikelets are often found to be more abundant than chaff, though in rare instances both are found in profusion. The incorporation of a large quantity of spikelets — an obviously useful stuff — by people recently emerged from the habit of food gathering remains inexplicable unless we assume that the food grains were produced in quantity, exceeding their consumption.

This practice, presumably originally unintentional, was studied to discover the stage at which it became intentional and to see if it could provide evidence of a change-over from wild rice to a domesticated form. Among other characters perhaps the presence or absence of awns could be fruitfully employed for distinguishing the wild from the cultivated rice. However, the awn has rarely been found preserved. Generally speaking the subfossil rice both in imprints and in the carbonised state has been shown to be without awns. Even any indication of the former presence of awns is lacking except in Neolithic husk from Baidipur, Orissa (studied by Vishnu-Mittre) and in carbonised material from Hastinapur, an Iron Age site (Chowdhury and Ghosh, 1954—5). It is, however, not possible from imperfectly preserved spikelets or grains to reconstruct mother plants or to determine whether they were erect or spreading with compact or lax panicles, the characters which distinguish *O. sativa* from *O. perennis*. Even the hairs on the spikelets are not usually preserved, though hair bases have been observed in Hastinapur material (Chowdhury and Ghosh, 1954—5). Fragmentary lemmas and paleae, and occasionally rachillas, have also been found preserved. In the absence of a detailed study of the corresponding parts of spikelets of various Indian species, it has not been possible to assess the value of these characters for specific determination.

It has, however, been possible to study the geographical groups, the *indica* and *japonica* varieties, distinguishing the slender and somewhat flattened grains of the group *indica* from the short and roundish grains of the group *japonica*. Within this broad classification have also been found intermediate shapes, as occur in these groups today. A comparison of the measurable characters of Chalcolithic rice from

---

* From personal observations of Dr S. Govindaswamy, of the Central Rice Research Institute, Cuttack.

Navdatoli-Maheshwar with those of primitive varieties of *indica* and *japonica* rice from various States from India to Japan revealed that on breadth and T/B ratio the grains were of the *indica* type, on L/B ratio they were intermediate, and on length they resembled *japonica* (Vishnu-Mittre, 1962). The model or standard for the two used here is after Himada (1956).

In the absence of the above morphological criteria to distinguish the spikelets and grains of wild rice from those of the cultivated one, the other course open is to look into the size statistics of modern and subfossil grains. Table 2 gives data regarding *Oryza perennis* and *Oryza sativa* var. *spontanea* as lent to me by Mr H.K. Hakin of the Botany Section, Central Rice Research Institute, Cuttack. The data from other species were determined by Dr H.P. Gupta, Senior Research Assistant, now Junior Scientific Officer, in my department.

Table 2 reveals that the $L/(B \times T)$ indices for the species investigated are as follows:

O. perennis	2.20, 2.21
O. officinalis	2.36
O. rufipogon	2.64
O. sativa var. *spontanea*	1.77, 1.79
O. sativa var. *japonica*	1.70
O. sativa var. *indica*	1.71

The $L/(B \times T)$ ratio can perhaps be used profitably to compare with similar ratios from the carbonised rice. Surprisingly, the ratio of the carbonised grains from various sites ranging from 2300 B.C. to A.D. 1500 was found to be 1.03−1.47 by Ghosh (1961). This wide difference may be due to shrinkage of grains after carbonisation. Considerable statistical work on well identified species must be carried out before any useful comparisons can be made. However, hitherto archaeobotanical material of rice from India has always been referred to the domesticated species *O. sativa*.

Table 3 presents all the Indian archaeobotanical records of rice in chronological order together with the dimensions. Their geographical distribution and dating are set out in Map 2. In the case of impressions it has not been possible to take all the three dimensions hence $L/(B \times T)$ ratio has not been calculated for them. The ratio $L/(B \times T)$ for the carbonised grains varies from 1.03−2.65. At the Neolithic site, Chirand, Bihar (Vishnu-Mittre, 1972) this ratio for the large carbonised grains is 1.72, approaching that of modern *Oryza sativa*, and for the carbonised slender grains it is 2.65, which is close to that of *O. rufipogon*, the wild rice. Before any conclusions are drawn from these comparisons, the reduction in length and increase in breadth owing to carbonisation ought to be taken into consideration. I have collected no positive data yet.

Records further suggest that there was a variety with awns in Hastinapur, dated about 500−300 B.C. Earlier than that indications of awns are noticed in material from the top levels of the Neolithic of Orissa. The latter approaches more *O. perennis*. In other records presence or indications of presence of awns has not been seen, either by me or by other investigators.

The oldest records of rice come from Chirand, Bihar as discussed under wheat, and Lothal and Rangpur in Saurashtra dated about 2300 B.C. (Ghosh, 1961). A

Table 2 *Size statistics of wild and cultivated rice*

Species	Length L (mm)			Breadth B (mm)			Thickness T (mm)			B/T	L/(B × T)
	Av.	Min.	Max.	Av.	Min.	Max.	Av.	Min.	Max.		
Oryza perennis											
Indian	8.12	7.41	8.83	2.29	2.08	2.49	1.61	1.40	1.82	1.42	2.21
Southeast Asian	8.13	8.08	8.18	2.28	2.01	2.54	1.60	1.50	1.70	1.42	2.20
O. officinalis	4.25	4.00	4.50	2.12	2.00	2.25	0.85	0.75	0.95	2.49	2.36
O. rufipogon	7.00	7.50	8.50	2.65	2.50	3.00	1.00	1.00	1.00	2.65	2.64
O. sativa var. spontanea											
Indian	8.82	7.99	9.65	2.55	2.10	2.99	1.95	1.72	2.17	1.30	1.77
Southeast Asian	8.70	8.06	9.35	2.65	2.27	3.03	1.83	1.55	2.10	1.44	1.79
O. sativa var. japonica	6.50	6.00	7.00	2.75	2.50	3.00	1.38	1.25	1.50	1.27	1.71
O. sativa var. indica	5.25	5.00	5.50	1.88	1.75	2.00	1.22	1.00	1.25	1.54	1.70

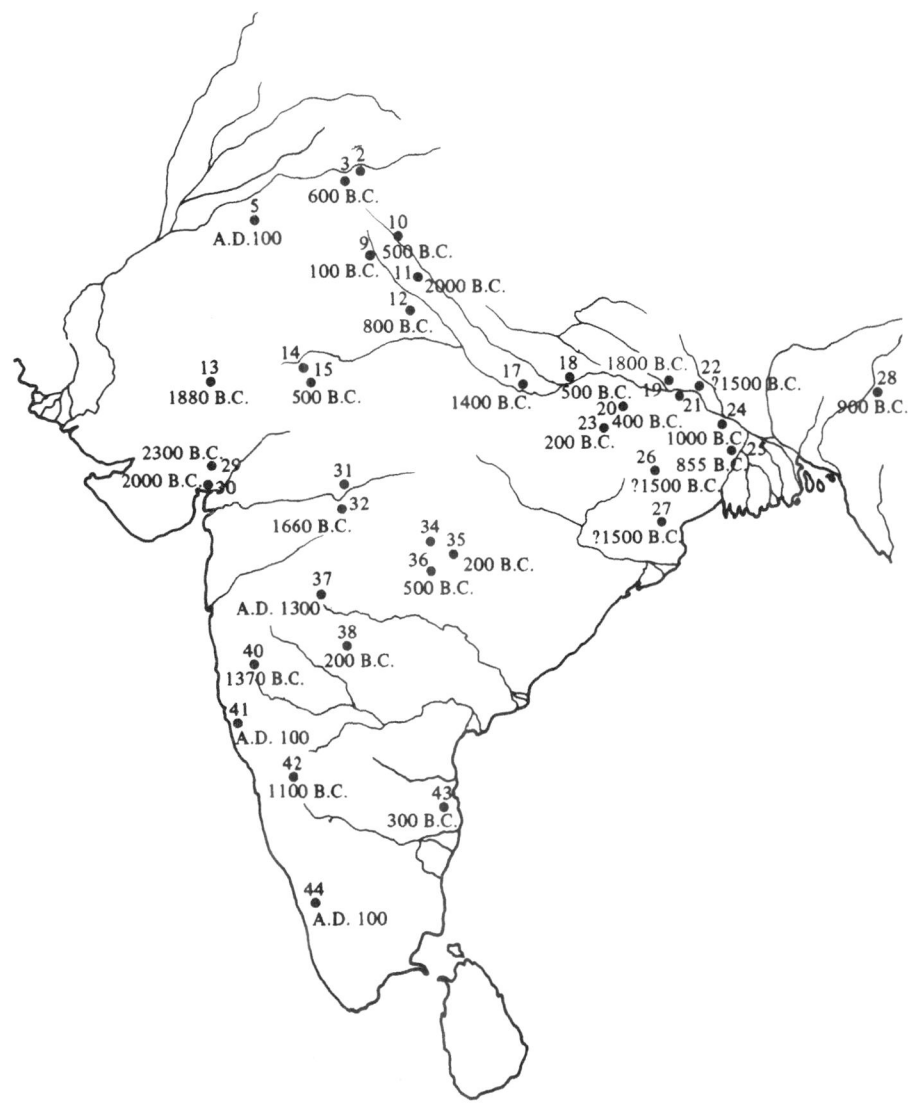

Map 2. Geographical distribution of records together with dates of rice in the Indian subcontinent. (For key see pp. 9–10)

few hundred years later, about 2000–1600 B.C., rice is now known to have been grown in the Gangetic plain at Atranji Khera in Uttar Pradesh (Chowdhury, Saraswat, Hasan and Gaur, 1971). About 1800 B.C. rice is again noticed in western India at Ahar in Rajasthan (Sankalia, Deo and Ansari, 1969; Vishnu-Mittre, 1969) and 200 years later about 1657–1400 B.C. at Navdatoli-Maheshwar in Madhya Pradesh.

The records from Orissa in eastern India are not dated but may be about 1500–1400 B.C. in a much more primitive context (Neolithic) than those found in

Table 3 *Rice records from India*

Key No.	Age	Site	Nature of material	Length L (mm)	Breadth B (mm)	Thickness T (mm)	$L/(B \times T)$
2		Kangra Fort (Punjab)	Charred grains	4.80–6.20 (5.49)	2.20–2.80 (2.45)	1.80–2.00 (1.96)	1.15
37	A.D. 1318–1759	Nevasa (Maharashtra)	Charred grains	3.94–4.74 (4.38)	1.96–2.46 (2.18)	1.28–1.75 (1.38)	1.47
44	Early Christian era	Pariyapuram (Kerala)	Husk	–	–	–	–
5	A.D. 100–400	Rangmahal (Rajasthan)	Husk	–	–	–	–
41	A.D. 100	Kolhapur (Maharashtra)	Charred grains	3.80–5.40	2.30–2.60	–	–
9	100 B.C.	Khokhrakot (Punjab)	Impressions	–	–	–	–
38	155 B.C.–A.D. 260	Ter (Maharashtra)	Charred grains and spikelets	3.00–5.00	2.50–2.75	1.00–2.25	1.30
				6.00–6.50	2.50–3.00	1.50–2.50	1.60
35	200 B.C.	Pauni (Maharashtra)	Impressions	6.00–7.00	2.25–2.50		
20	200 B.C.	Rajgir (Bihar)		Data not available			
43	300 B.C.	Kunnatur (Madras)	Spikelets	7.00	2.00	–	–
36	500–400 B.C. A.D. 300–600	Paunar (Maharashtra)	Spikelets	5.00–9.00	1.00–3.00	–	–
21	405–115 B.C.	Pataliputra (Bihar)	Charred spikelets and grains	3.50–5.40 (4.95)	2.50–3.00 (2.66)	2.00–2.50 (1.82)	1.03
34	500–200 B.C.	Kaundinyapur (Madhya Pradesh)	Charred grains	3.00–4.80 (3.90)	1.50–3.00 (2.25)	1.00–1.50 (1.25)	1.37
14	500–200 B.C.	Nagda Ujjain (Madhya Pradesh)	Charred spikelets and grains	4.50–5.70 (4.90)	2.10–2.60 (2.47)	1.50–2.00 (1.78)	1.76
15	500 B.C.	Garh Kalika (Madhya Pradesh)	Charred spikelets and grains	4.00–5.50 (4.73)	2.10–2.70 (2.49)	1.00–1.70 (1.48)	1.28
18	500 B.C.	Rajghat (U.P.)		Data not available			
10	506–306 B.C.	Hastinapur (U.P.)	Charred spikelets	5.30–7.00	2.00–2.70	–	–

3	600–200 B.C.	Rupar (Punjab)	Charred grains	—	—	—	—
23	637 B.C.	Sonpur (Bihar)	Charred grains	4.16–5.54 (4.85)	1.65–2.57 (2.11)	1.40–1.60 (1.50)	1.33
12	821–604 B.C.	Noh (Rajasthan)					
28	900 B.C.	Ambri (Assam)	Impressions	Data not available			
42	870 B.C.	Hallur (Mysore)	Impressions	7.00–8.00	2.00–2.50	—	—
25	1000 B.C.	Pandu Rajar Dhibi (Bengal)		Data not available			
24	1385–1085 B.C.	Mahesdal (Bengal)	Charred grains and impressions				
40	1370 B.C.	Inamgaon (Maharashtra)					
17	1400 B.C.	Kausambhi (U.P.)					
32,31	1557–1400 B.C.	Navdatoli-Maheshwar (Madhya Pradesh)	Charred grains	4.30–5.10 (4.70)	2.00–2.40 (2.20)	1.20–1.80 (1.50)	1.42
13	1885–1070 B.C.	Ahar (Rajasthan)	Impressions	5.00–7.00 (6.00)	2.00–3.00 (2.50)	1.25–2.00 (1.63)	1.47
30	2000–1800 B.C.	Rangpur (Saurashtra)	Impressions	Data not available			
11	2000–200 B.C.	Atranji Khera (U.P.)	Charred grains	5.00–7.00	2.50–3.00	—	—
29	2300 B.C.	Lothal (Saurashtra)	Impressions (only 3)	Data not available			
27	Late Neolithic	Baidipur (Orissa)	Husks	5.00–5.50 (5.25)	2.25–2.50 (2.38)	1.25–2.00 (1.63)	1.35
26	Neolithic	Singhbhum (Bihar)	Charred grains	5.00–5.25 (5.12)	2.00–2.25 (2.12)	1.25–1.50 (1.35)	1.76
22	Neolithic	Oriyup (Bihar)	Impressions	5.00–5.50	2.25–2.50	—	—
19	2500–1800 B.C. (Neolithic)	Chirand (Bihar)	Charred grains — Small	4.25–4.50 (4.38)	1.50–1.80 (1.65)	1.00–1.00 (1.00)	2.65
			Large	5.00–5.25 (5.13)	2.25–2.50 (2.38)	1.25–1.25 (1.25)	1.72

highly evolved late-Harappan and Chalcolithic cultures in western India (Vishnu-Mittre, 1968c).

The earliest records from south India are from the Iron Age (Hallur: 870 B.C., Vishnu-Mittre, 1971a).

The discovery of Neolithic rice from Bihar and a few radiocarbon dates so far obtained throw some doubt on the thesis of Agrawal (1969) that rice was domesticated in western India and subsequently diffused gradually into the rest of India. It is hoped that the new discoveries, such as from Chirand, and more radiocarbon dates will provide an independent check on the conclusions arrived at through cytogenetical researches and through the distribution of progenitors of rice which are concentrated in the east.

Amongst the records of rice from India, those discovered from Kaundinyapur (Vishnu-Mittre, 1966a; Vishnu-Mittre, 1968d) comprise the smallest grains but their average $L/(B \times T)$ is 1.37. In view of the intermediate forms occurring between the subgroup *indica* and *japonica*, it is not advisable to refer them to the *japonica* group.

Of all the sites which have yielded archaeobotanical records, in only three – Ahar, Navdatoli-Maheshwar and Atranji Khera – have continuous records over a period of 800 years been studied. In the remaining sites there are occasional records only. Surprisingly no appreciable change in morphology and size of rice spikelets or grains has been noted which would tell us of breed or breeds different from the others.

For rice cultivation about 4000 years ago in Saurashtra and Rajasthan, a regime of higher rainfall than today in these regions has been opined by Ghosh (1961). Rice was unknown during the Harappan civilisation (2300–1750 B.C.) in the Indus valley. Was the rainfall regime different here? Further, a high rainfall regime in Rajasthan, as inferred from the finds of rice there, would be inimical to the cultivation of barley at Kalibangan, where abundant barley material has been found.

Barley

A few grains of barley have been found mixed with wheat at Mohenjo-daro and identified by Luthra (1936) as *Hordeum vulgare* var. *nudum.* In low frequency they are also found mixed with wheat at Chanhu-daro (Vishnu-Mittre, unpublished). The material from Harappa is referred to *H. vulgare* var. *hexastichum* (Vats, 1940). In contrast to the low frequency of barley found at the above sites, a sizeable quantity was found at Kalibangan (Vishnu-Mittre, unpublished). Mohenjo-daro material as mentioned above is dated 1750 B.C. from the late levels, whereas Kalibangan material is dated to 2090–2075 B.C. (Agrawal and Kusumgar, 1968a, b). There is reason to believe that barley was used throughout the Harappan period from 2300 B.C. to 1750 B.C.

The Kalibangan material consists of small and large grains and some of them are twisted, indicating that they were produced in two lateral rows and suggesting that they belonged to six-rowed barley. The collection includes both naked and hulled forms. The former are recognised by the characteristic transverse rippling of the grain shell. Such grains in cross-section look rounded as contrasted from the angular outline of the hulled forms. Internodes are absent and it is not possible to determine whether the spikes were erect or nodding. The grains are much shrunk in size (naked,

3.0–5.0 × 1.75–3.0 × 1.25–3.0 mm and hulled, 4.0–5.0 × 1.75–3.5 × 2.0–3.0 mm). Their small size may be due partly to growth on poor sandy soils and partly to carbonisation. The latter can cause a considerable decrease in length and an increase in thickness.

More or less contemporary with the Harappan civilisation, records of barley have recently been discovered from the Gangetic plain, from Atranji Khera in U.P. dated to 2000–1500 B.C. (Chowdhury, Saraswat, Hasan and Gaur, 1971) and from the Neolithic of Chirand, Bihar estimated to date from 2500–1800 B.C. or earlier (Narain, 1970; Vishnu-Mittre, 1971c, 1972). The Neolithic material from this site comprises both naked and hulled forms, the former outnumbering the latter. Some of the grains show slight deviation from the straight shape, suggesting their derivation from a six-row type. The barley grains measure about 4.0–5.0 × 2.25–3.25 × 1.50–2.25 mm and have been referred to *Hordeum* sp. (Vishnu-Mittre, 1972).

Chalcolithic, Iron age and Early Historical records are known from Atranji Khera dated 1200–600 B.C. (Buth and Chowdhury, 1971) and subsequently barley is found at Ter, Osmanabad, Maharashtra around 155 B.C. to A.D. 160 (Vishnu-Mittre *et al.* 1972). Only hulled barley, about 5.5 × 2.0 × 1.55 mm, was found at the latter site.

The distribution of finds of barley are set out in Map 1.

The millets
Identification of the seeds of the millets depends primarily on rather small differences in shape. For comparative purposes these may be set out as follows:

Eleusine coracana:	Finger millet; Ragi:	globose
Eleusine indica:	Wild Ragi:	oblong
Sorghum bicolor:	Sorghum; Jowar:	ovoid, subrotund to orbicular or elliptic oblong
Pennisetum typhoideum:	Pearl millet; Bajra:	one end broader than the other
Pennisetum typhoideum: *Panicum miliaceum:*	Italian millet:	spindle-shaped and biconvex
Paspalum scrobiculatum:		spindle-shaped to plano-convex

In addition, the position and size of the hilum scar is a valuable aid to classification. In general, sorghum grains are larger than grains of the other millets, but the effect of carbonisation is uncertain, and there is the possibility of changes in size as a result of recent breeding work, so size alone is of doubtful diagnostic value.

Three of the millets, finger millet, sorghum and pearl millet, are regarded on botanical grounds as of African origin. Archaeological evidence of their antiquity and distribution in India is therefore of particular importance.

The distribution of finds of millets is set out in Map 3.

Finger millet; Ragi
Carbonised grains of Ragi, *Eleusine coracana,* are known from Mysore State from the Neolithic of Hallur dated about 1800 B.C. (Vishnu-Mittre, 1971a). In the

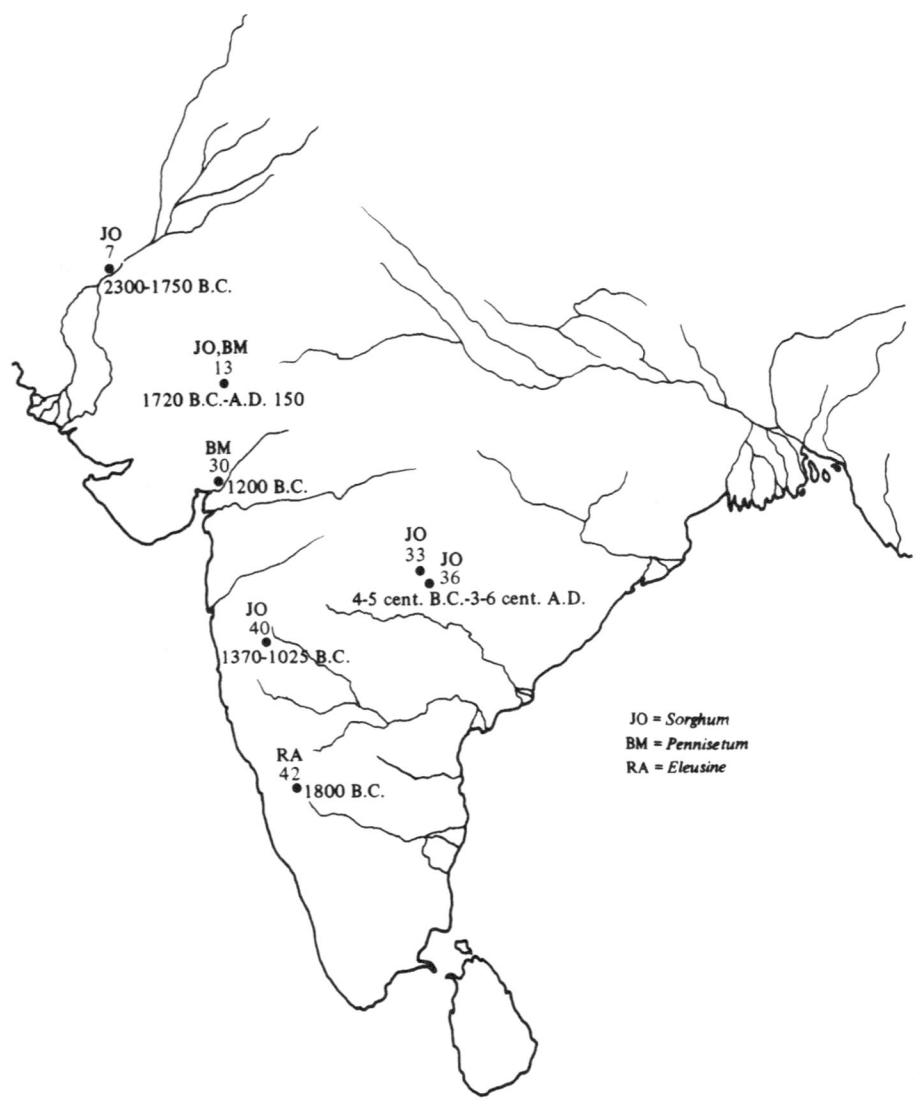

JO
7
2300-1750 B.C.

JO,BM
13
1720 B.C.-A.D. 150

BM
30
1200 B.C.

JO
33   JO
     36
4-5 cent. B.C.-3-6 cent. A.D.

JO
40
1370-1025 B.C.

RA
42
1800 B.C.

JO = *Sorghum*
BM = *Pennisetum*
RA = *Eleusine*

Map 3. Geographical distribution of records together with dates of millets in the Indian subcontinent. (For key see pp. 9–10.)

charred material from this site both oblong and globose grains have been found, one each in a spikelet (Table 4). The small number released from the fragile charred mass consisted of more oblong grains than globose ones. The globose grains compare with those of *Eleusine coracana* and oblong ones with those of *E. indica*. The spikelets in the charred mass appear to have been derived from both species. The Neolithic context of the material and its date, 1800 B.C., and the discovery of Ragi together with its relative, *E. indica*, are suggestive of an early stage in the history of the crop.

Table 4 *Dimensions of* Eleusine coracana *and* E. indica *from Hallur (averages in brackets)*

Oblong (*Eleusine indica*)		Globose (*Eleusine coracana*)	
Length *L* (mm)	Breadth *B* (mm)	Length *L* (mm)	Breadth *B* (mm)
1.50–2.00	0.75–1.00	1.50–1.00	1.50–1.00
(1.80)	(0.90)	(1.18)	(1.18)

Evidence adduced by Mehra (1963*a, b*) for the view that the wild *E. indica* is a diploid of Indian origin and *E. coracana* a tetraploid of African provenance poses problems on which the archaeological evidence so far throws no light.

The recent report of Ragi from Paiyampalli, Madras (Allchin, 1968) and dated to about 645 B.C. is misleading. The material kindly sent to me by the excavator of the site shows nothing but a legume.

Sorghum; Jowar

A drawing on a potsherd from Mohenjo-daro (Marshall, 1931, Plate LXXXVII, Photo 5) interpreted as that of sorghum is so far the only record from the Indus Valley civilisation. I have not had the occasion to examine either the specimen or the photo, but this evidence alone should not be taken very seriously. Obviously its history during the period 2300 B.C. to 1750 B.C. is doubtful.

Subsequent to the end of the Harappan civilisation records of sorghum are known from Ahar, Rajasthan dated about 1725 ± 110 B.C., but more profusely from Period I and Phase I*c* dated about 1550 ± 110 to 1270 ± 110 B.C. (Sankalia *et al.* 1969; Vishnu-Mittre, 1969). Regarding the latter, the excavators caution us owing to the disturbed nature of the top layers of this phase. In view of its occurrence at this site a few centuries earlier, there seems to be no reason to doubt its occurrence in I*c,* despite the evidence that the top of I*c* was deserted and recolonised by the Iron Age and Early Historical people with a considerable gap in time. Profuse records are however noted in Period II, Phase II*b* corresponding to the 'Kushan Period' about 150 B.C. to A.D. 50.

Records of sorghum from this site are all found in the form of impressions on potsherds or within their matrix as a result of the practice of mixing grain and chaff of this cereal with clay before turning it into pots. The imprints or the embedded grains are ovoid in shape, subrotund to orbicular or elliptic oblong and slightly compressed, more or less round but slightly pointed on one of the ends. In some the ribbed palaea has been reduced to ash without any anatomical details preserved. The Ahar sorghum has been found to compare with quite a few modern Indian cultivars. The small size of Ahar material (2.0–4.0 × 2.0–3.0 × 1.0–1.5 mm) may be due to shrinkage during preservation.

The other two records of sorghum dated to the 4–5th century B.C. to the 3–6th century A.D. are from Maharashtra. Of these, records from Paunar, Maharashtra are based on impressions and range in size from 2.3–3.0 × 1.3–2.5 mm (Vishnu-Mittre and Gupta, 1968). Carbonised sorghum from Bhat Kuli, dist. Amraoti, Maharashtra ranges from 2.3–4.0 × 2.0–3.0 mm.

Pearl millet; Bajra

The earliest known record of *Pennisetum typhoideum* is from Rangpur Period III in Saurashtra dated about 1200–1000 B.C. (conservative estimate based on archaeological dating). The obovate carbonised grains recovered from the charred mass measure 3.0–4.0 X 2.5–3.0 mm and have been recognised as those of pearl millet by Ghosh and Lal (1962–3). Some of the impressions from Ahar also seem to belong to pearl millet as they differ in shape from those of sorghum (Vishnu-Mittre, 1969).

Paspalum scrobiculatum

A few rotund, concavo-convex spikelets measuring about 1.0 X 1.5–2.0 mm in the incrustations on iron arrow heads from Hallur, Mysore and dated about 1000 B.C. are doubtfully referred to *P. scrobiculatum* (Vishnu-Mittre, 1968c, 1971a). A thousand years later at Ter in Maharashtra a carbonised grain measuring about 2.25 X 1.75 X 1.25 mm has been discovered (Vishnu-Mittre *et al.* 1972). Some of the impressions from Ahar in Rajasthan might as well belong to this species, as their shape and dimensions are more or less similar. The identification of the above two millets may have to be revised if and when more material has been discovered and the identity supported by comparative data from the extant species.

Maize

There is, as yet, not much evidence regarding the history of *Zea mays* in India. Pollen evidence from the Kashmir Valley (Vishnu-Mittre and Gupta, 1966) is dated by conservative estimate to the fifteenth century A.D. Impressions on a potsherd from Kaundinyapur, Madhya Pradesh and dated archaeologically to about A.D. 1435 have created a difference of opinion among the experts, namely Manglesdorf, Jeffreys and Carter to cite a few (Vishnu-Mittre and Gupta, 1966; Vishnu-Mittre, 1966a, 1968d). That the impressions could be of a piece of basketry, or of course textile or fabric as opined by Dr Richard McNeish of the National Museum of Canada and Dr James Griffin of the University of Michigan has been disproved by the experiments conducted by Dr George F. Carter (personal communication) of the Isaiah Bowman Department of Geography, Johns Hopkins University, Baltimore, Maryland. It appears to me that the controversy involves the adherance of either of the schools to their views of whether maize reached Asia and Africa in the pre- or post-Columbian era. This discovery tends to support a pre-Columbian introduction. In any event, as opined by Jeffreys (1965), the credit goes to the Arabs for introducing it into India rather than to the Portuguese. The supporting evidence for this is provided by Indian names for maize, namely: *Makka jauri* (Mecca sorghum), *Mekka jola* (Mecca sorghum), *Makkai* (grain of Mecca), *Mokka jona* (Mecca sorghum), *Mukka cholam* (Mecca sorghum), etc.

Apart from the problem of its introduction, the enigmatic living fossil maize in Sikkim perhaps tends to suggest a reconsideration of the occurrence of maize there. Several distinctive varieties of it are grown by the aborigines in Siam, Burma, Assam, Sikkim, China, Tibet, etc. It seems to have been grown there for a long time. It can hardly have been introduced by the Arabs through the inland caravan routes. The local names used for maize by the Nagas (*Konoma tsunghundhro, Wokotsu*), the

Lushias (*Pawi Vaimin, Sap Vaimin, Bangpu Vaimin*), the Kukis (*Kolbu*) and the Khasis (*Riew Haddem*) do not indicate its diffusion through the Arabs but either from the neighbouring village or from Burma or China; even some names refer to it as European maize or foreign maize (*Bilati Hangi*) among the Chang Nagas. The varieties of names suggests a rather long period of cultivation and that it was an introduced crop.

Legumes

Horse gram

Referred to *Dolichos lablab* by the excavator of the site (Nagaraja Rao and Malhotra, 1965), the carbonised material from the Neolithic site, Tekkalkota, in Mysore State has been identified by me as *Dolichos biflorus* (Vishnu-Mittre, 1968*b, c*). The oldest record at the site goes to 1780–1540 B.C. (Agrawal and Kusumgar, 1966). In general form the specimens also compare with the seeds of *Glycine max* but in shape they are more like those of *D. biflorus,* though some are not so flattened as in this species. I attribute this to carbonisation. Similar seeds are found at this site in the ash pit dated to 303 B.C.

Seeds or cotyledons also occur scattered in several samples from Ter, Maharashtra, an early historical site (Vishnu-Mittre *et al.* 1972). Specimens from Ter are larger than those from Tekkalkota, as shown in Table 5. The dimensions are of interest. While there is no change in size even after 1400 years at Tekkalkota, those from Ter about 100 years later and at a distance of about 100 km to the north are 1½ times longer and slightly broader.

Peas

The record of peas (*Pisum arvense*) from the Harappan civilisation is from Harappa itself (Vats, 1940). Carbonised seeds have recently been discovered from Chirand, the Neolithic site in Bihar (Vishnu-Mittre, 1971*c,* 1972). A few scattered seeds were found mixed with wheat and pulses in several samples from Navdatoli-Maheshwar, a Chalcolithic site, in Madhya Pradesh (Vishnu-Mittre, 1962). The early historical records occur at Khokhrakot, Rohtak (Sahni, 1936, 1938) and at Ter, Osmanabad, Maharashtra. A sizeable collection has been discovered from Ter only. Such measurements as are available are given in Table 6.

Chick-pea

The oldest record of chick-pea (*Cicer arietinum*) is from Atranji Khera in U.P. and is believed to date from about 2000 B.C. (thermoluminescent dating) (Chowdhury, *et al.* 1971). The early historical records are from Maharashtra only (Ter, dist. Osmanabad, and Bhatkuli, dist. Amraoti). These are dated at about 150 B.C. to A.D. 200 (Vishnu-Mittre, 1968*c;* Vishnu-Mittre and Gupta, 1968–9; Vishnu-Mittre, *et al.* 1972). The seeds measure 2.3–5.0 × 2.0–4.0 × 2.0–4.0 mm. A record dated about the third to fifth century B.C. from Nevasa (Sankalia, Deo and Ansari, 1960) remains to be confirmed. The record from Navdatoli-Maheshwar (Sankalia, Subbarao and Deo, 1958), is based on a wrong identification. So far the records are very late in time. Its Sanskrit name *Chennuka* from which modern Indian names are

Table 5   *Dimensions of* Dolichos biflorus *seeds from Indian sites*

Age	Site	Length L (mm)	Breadth B (mm)	Thickness T (mm)
150 B.C.–A.D. 100	Ter	6–9	4–5	2–3
330 B.C.	Tekkalkota	4–6	3–4	2–2.5
1780–1540 B.C.	Tekkalkota	4–6	3–4	2–2.5

Table 6   *Dimensions of* Pisum arvense *from Indian sites*

Age	Site	Size (mm)
100 B.C.	Khokhrakat, Punjab	–
150 B.C.–A.D. 100	Ter, Maharashtra	2.5–4.0 × 2.5–4.5
1658–1443 B.C.	Navdatoli-Mahashwar,	2.0–4.0 × 2.0–4.0
2100–1850 B.C.	Harappa, Punjab	
2500–1800 B.C.	* Chirand, Bihar	2.0–4.0 × 2.0–4.0

* The smaller grains at this site measure 2.0 × 1.0–1.25 mm, and these may be either abortive grains of *Pisum arvense* or grains of some other legume (Vishnu-Mittre, 1972)

derived may suggest its antiquity at least to 1100 B.C., since the authors of the Painted Grey Ware are believed to be Aryans. Chick-pea as food for horses has been in use in India south of the Vindhyan mountains since about A.D. 800. North of the Vindhyas barley was fed (Gode, 1945, 1946). Since India is regarded as one of the centres of origin of chick-pea (Vavilov, 1951), we are sure to get many older records here. Material of chick-pea from Atranji Khera in U.P. indeed extends its history in India to about 2000 B.C.

Phaseolus spp:
Hitherto the oldest records of both *P. mungo* and *P. aureus* are from the Chalcolithic site, Navdatoli-Maheshwar dated to 1660–1440 B.C. (Vishnu-Mittre, 1962, 1968c). The seeds of both have been identified from the shape and size of carbonised grains (*P. mungo:* square to oblong, 3.5–4.7 mm long, 2.5–3.0 mm broad and 0.5–0.75 mm thick; *P. aureus:* square to roundish, 2.5–3.0 mm long, 1.8–2.2 mm broad and 1.5–2.0 mm thick). Carbonised seeds measuring 3.0–4.0 × 1.5–2.0 × 1.5–2.0 mm from Paiyampalli, Madras, a Neolithic/Megalithic site dated about 645 B.C., are more or less similar to those of *P. mungo* but are less broad and thicker than the Navdatoli grains and the hilum scar is not exactly placed in the middle (Vishnu-Mittre). Grains from Ter, 150 B.C. to A.D., 100 measure 3.25–4.0 × 3.3 × 0.75–1.5 mm. India together with the central Asiatic centre is believed to be the home of these two species (Vavilov, 1951).

Lentils
The earliest record of *Lens culinaris* has been recently discovered at the Neolithic site Chirand, Bihar (Vishnu-Mittre, 1972). Among the later records there are two, the one from the Chalcolithic site, Navdatoli-Maheshwar (Vishnu-Mittre, 1962,

24

Table 7   *Dimensions of lentil grains from Indian sites*

Age	Site	Dimensions
150 B.C. to A.D. 100	Ter	3.4–5.0 × 1.3–2.5
1550 to 1440 B.C.	Navdatoli-Maheshwar	2.2–3.1 × 1.8–2.0
2500 to 1800 B.C.	Chirand	2.25–3.0 × 0.8–1.25

1968*c*) and the other from the early historical site, Ter, Osmanabad, Maharashtra (Vishnu-Mittre *et al.* 1972). The grains of Ter are larger in size than those from other sites (Table 7).

Lathyrus sativus
The earliest record of the wedge-shaped and compressed seeds of this pulse comes from the Neolithic site Chirand, Bihar (Vishnu-Mittre, 1972). It is also known from Navdatoli-Maheshwar about 1660 to 1440 B.C., and subsequently from the Early Historical period at Ter, Maharashtra and Kaundinyapur, Madhya Pradesh (Vishnu-Mittre, 1962, 1966*a*, 1968*c*, *d*; Vishnu-Mittre *et al.* 1972). The seeds are of variable size (3.0–4.5 × 3.0–3.5 × 2.0–3.0 mm). Carbonised seeds referred to *Lathyrus* spp. have been reported from Atranji Khera dated to about 2000–1500 B.C. (Chowdhury, *et al.* 1971).

Leguminous weeds
Seeds of *Medicago denticulata, M. falcata* and *Lotus corniculatus* are known from the Neolithic of Kashmir from Burzahom, 2300–1500 B.C. (Vishnu-Mittre, 1968*a*, *c*). Carbonised seeds of *Lathyrus sphaericus, Vicia sativa, V. tetrasperma* are known from Navdatoli-Maheshwar, 1660 B.C. (Vishnu-Mittre, 1962, 1968*c*). *Lathyrus sphaericus* is also known from the early historical site at Kaundinyapur, Madhya Pradesh (Vishnu-Mittre, 1966*a*, 1968*c*, *d*).

Oil seeds
The records of oil seeds are scattered. Remains of *Sesamum indicum* are known from Harappa (Mackay, 1943; Vats, 1940). Seeds of *Brassica juncea* reported from Chanhu-daro by Mackay (1943) have not been noted in the material examined by me.

Carbonised seeds of *Linum usitatissimum* and myrobalan (*Phyllanthus emblica*) are known from 1600 to 1440 B.C. at Navdatoli–Maheshwar (Vishnu-Mittre, 1962). A carbonised seed of *Ricinus communis* is known from Ter, 150 B.C. to A.D. 100 (Vishnu-Mittre, 1968*c*; Vishnu-Mittre *et al.* 1972).

Fibre plants
Cotton is known from Mohenjo-daro and Harappa from 2300 to 1750 B.C., as inferred both from actual fibres in contact with copper tools and a silver vessel and from impressions of textile (Marshall, 1931; Mackay, 1943; Vats, 1940). Cotton is also known from Nevasa about 1·500 B.C., together with silk (Gulati, 1961). Textile impressions have also been noted on Iron Age sherds from Mysore dated about 1000 B.C. (Allchin, 1968).

The evidence of the use of linseed as flax comes from the discovery of spun fibres at Chandoli, 1400–1200 B.C. (Gulati, 1965). Buth and Chowdhury (1971) have recently reported fibres of urticaceous plants from Atranji Khera, Phase III dated to *c*1200–600 B.C.

Fruits

Date palm

Two tiny faience sealings shaped like a date seed suggest familiarity of the ancient Indians at Harappa with the date palm (Vats, 1940). Matting impressions at Tekkalkota, Mysore and charcoal at Utnur (Allchin, 1968), both Neolithic sites, suggest the widespread use of date palm. A carbonised seed of date is known from the Chalcolithic site at Inamgaon, Maharashtra, some levels of which have been dated to 1370–1025 B.C. (Vishnu-Mittre, unpublished).

Melon

A few vestiges of seeds comparable to those of melon are known from Harappa from the porous earth of earthenware. The seeds were too brittle to be lifted up (Vats, 1940).

A representation of lotus fruit in faience and some earthenware vases shaped like a pomegranate and coconut fruits tend to suggest the familiarity of ancient Harappans with these fruits. Lemon cannot be left out of the question here since a well-made pendant shaped like a lemon leaf is suggestive of its presence (Vats, 1940).

Zizyphus nummularia

Carbonised stones have been recovered from Navdatoli (1660–1400 B.C.) and from Kaundinyapur and Ter (Vishnu-Mittre, 1962, 1966*a*, 1968*c*, *d*; Vishnu-Mittre *et al.* 1972). They have also been found at Inamgaon, Maharashtra (Vishnu-Mittre *et al.*, unpublished).

Phyllanthus emblica

Carbonised seeds dated 1600 B.C. have been found at Navdatoli (Vishnu-Mittre, 1962).

Saccharum spp.

Charcoals of *Saccharum arundinaceum* from 2300 B.C. to 1750 B.C. have been recovered from Harappa (Vats, 1940). Remains of *Saccharum* probably belonging to *spontaneum* from Chirand, Bihar and dated about 1800 B.C. are under investigation by me. *S. spontaneum* from Hastinapur has been recognised by Chowdhury and Ghosh (1955). The material from the latter two sites is mixed with clay. *S. arundinaceum* and *S. spontaneum* are not sugary. They were probably used for basketry and matting. There is no information so far regarding *S. officinarum*, the tropical sugarcane.

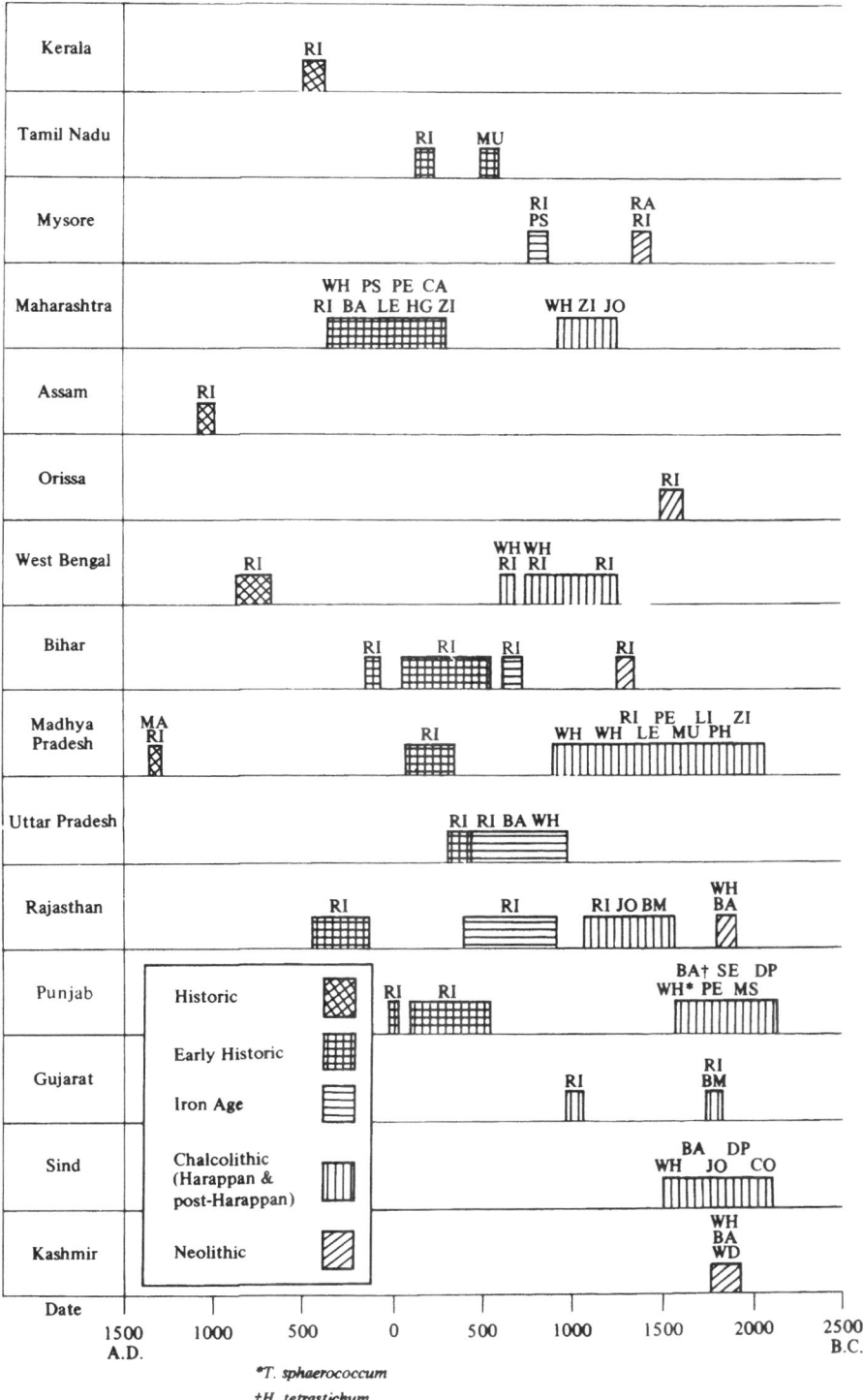

Fig. 1. Distribution of cultivated plants in space and time in the Indian subcontinent. (For key see p. 10.)

Conclusion

In Fig. 1 crop records are set out by state against the cultural sequence and the approximate chronology. The information on the ancient Indian crops which has accumulated during the last decade through pollen-analytical and archaeobotanical researches lacks the phases of interaction between natural and human selection. In archaeological terms it has not yet been possible to discover the early cultivars, or the progenitors of present-day crops and the circumstances under which they were domesticated. We have nevertheless discovered weeds in the Neolithic sites at Burzahom (Kashmir), Chirand (Bihar) and Hallur (Mysore) suggesting that the food-gathering habit continued into the Neolithic. This habit was not restricted to the Neolithic peoples as has become apparent from the investigations (in progress) by Miss R. Savithri and myself of carbonised material from the recently discovered Harappan site, Surkotada about 10 km northeast of Adesar in district Kutch, Gujarat (Joshi, 1972). The material consists of as many as thirty-four kinds of seeds tentatively identified as: *Scirpus supinus,* species of *Carex, Eriophorum* and other sedges; *Setaria intermedia, Eleusine indica, Phragmites karka* and other grasses; *Atriplex stocksii, Amaranthus* sp.; *Euphorbia* sp. cf. *pycnostegia;* and *Polygonum* spp., etc. The question obviously arises, were these Harappans, the food gatherers, the builders of fortified citadels at this site? We continue to be ignorant of the effects of hybridisation between primitive and advanced races or between cultivars and their wild relatives, and further, howsoever copious and valuable the information presented here may be, we have failed to record changes which human selection might have created. For instance, *Triticum sphaerococcum* and *Oryza sativa* are products of the plant breeding skill of ancient Indians, but how it was achieved we are yet unable to say. Even the earliest ancient crops are developed forms, suggesting that we have not yet struck the earliest strata or else the sites with remains of domestication have yet to be discovered.

However, the history of several of our cultivated plants has been traced. Wheat, barley and rice are the oldest cereals in the country, dating from 2500 B.C. Ragi, together with its relative, dates from 1800 B.C., *Sorghum* is equally ancient, contrary to the view of Burkill (1953) that it was introduced into India after A.D. 700 by the Yemenite Arabs from Zanzibar. Pearl millet dates from 1200 B.C., although *Paspalum scrobiculatum* is later. Evidence presented suggests reconciliation with the view that maize was introduced into India in the pre-Columbian era and probably by the Arabs — a case of transatlantic diffusion, though the source of the enigmatic maize of Sikkim and Assam remains unknown.

Pisum arvense, lentil and *Lathyrus sativus* are the oldest legumes to have been used in India. The chick-pea is the next oldest, dating from about 2000 B.C. *Dolichos biflorus* has been in use since 1780 B.C. and *Phaseolus mungo* and *P. aureus* since 1660 B.C.

Mustard and sesame are the oldest oil seeds, whereas castor seed occurs much later in history. Linseed dates from 1600 B.C. Cotton is dated 2300 B.C.

The date is probably the oldest fruit with which the ancient Indians were familiar. *Zizyphus nummularia* and *Phyllanthus emblica* date from 1660 B.C.

Comparative studies of the food grains discovered from the various sites within the last 4500 years do not bring out much information regarding the evolution of new strains in the cereals but some of the legumes such as *Dolichos biflorus, Phaseolus mungo* and *P. aureus* do, however, show slight increases in seed size, especially towards the commencement of the Christian era.

India belongs to two of Vavilov's (1951) centres of origin of cultivated plants, namely the Hindustan centre and the central Asiatic centre. Hitherto the oldest records of most of these cultivated plants have been in the respective centres of their postulated origin. The oldest record of rice found in western India in Saurashtra and the other radiocarbon-dated archaeological records of rice led Agrawal (1969) to suggest that, after its domestication in western India, rice was gradually diffused into the south and east of India. Recent excavations in Bihar have revealed Neolithic rice at Oriyup and at Chirand. The latter site is radiocarbon dated to 1800 B.C. Fresh material from Chirand newly assayed for ^{14}C date at the Tata Institute of Fundamental Research shows the top layers (dated 1500–1800 B.C.) older than the bottom layers in which rice has been found (Agrawal, 1971), obviously suggesting that the Neolithic rice from Chirand is indeed much older. Narain (1970) estimates its age as *c.* 2500 B.C. As discussed elsewhere in this text I have estimated its age to be around 3500 B.C. The evidence as presented does not support the thesis of Agrawal (1969) that rice was domesticated in western India. The suggestion is put forward in this volume (p. 58) that rice may have been domesticated over a very broad area in south and southeast Asia where its wild progenitor is indigenous. The archaeological evidence does support this theory, but it will be necessary to await more comprehensive excavations in eastern areas, and adequate radiocarbon dating. The present time is therefore inopportune for drawing conclusions.

The oldest records of wheat, *T. sphaerococcum,* in India tend to suggest its origin there. It has subsequently been diffused into Madhya Pradesh and a few centuries later into Maharashtra. It was not until the Iron Age that it was diffused into Bengal. Its diffusion into central India and Maharashtra did not take place through Rajasthan.

The oldest records in India of barley, of which the centre of origin is outside India, are concentrated in the north and the northwest. If was diffused into Maharashtra towards the beginning of the Christian era.

The Indian records of *Sorghum, Pennisetum* and *Eleusine coracana* are the oldest known in the world. Botanically the primary centres for their origin are believed to be in Africa. The extensive differentation that has taken place in India justifies the regarding of India as a secondary centre for these cultivars. *Sorghum,* the oldest millet in the country, has been gradually diffused through Rajasthan into Maharashtra. The oldest record of *Pennisetum* is dated to 1200 B.C. and is from Saurashtra. *Paspalum scrobiculatum,* of which India is the centre of origin, dates from the Iron Age.

Evidence of cultural contacts of the ancient Indians with Africa is meagre. Indications of trade contacts with Egypt require definite establishment. The discovery of the remarkable brick dockyard connected by a channel to the gulf of Cambay at Lothal in Saurashtra has yet to be evaluated for any contacts of the Harappans with Africa. The clustering of records of millets along the western coast may be suggestive of their diffusion through maritime trade. Supporting evidence has yet to be found.

A few sites have yielded enough samples from each cultural period to enable us to infer local changes in the pattern of food economy at these sites. The Chalcolithic site Ahar in Rajasthan covering a span of time from 1940 ± 95 B.C. to A.D. 1750 ± 95 reveals rice to have been the only cereal until about 1725 ± 110 B.C. when *Sorghum* was first introduced. It was not until the beginning of the Iron Age that *Sorghum* and other millets assumed importance in the staple diet of the ancient Aharians. At Navdatoli-Maheshwar, a Chalcolithic site in Madhya Pradesh, wheat was the only cereal until rice was introduced in Period II and gradually grew in importance in 1440 B.C. The exact time of the introduction of rice cannot be determined because of erratic radiocarbon dates. The same site also reveals increased use of linseed and *Lathyrus sativus* in 1440 B.C., whereas lentil and *Pisum arvense* had declined by 1440 B.C. At Ter in Maharashtra during the Satavannah period (200 B.C. to A.D. 100) wheat and rice were equally important but in the late Satavannah period (A.D. 100–250) wheat predominated, the records of rice being just half those of wheat. Chick-pea made its first appearance during late Satavannah period and became an important article of food economy in the post-Satavannah period (A.D. 250–400) when barley and *Paspalum scrobiculatum* also entered the food economy comprising wheat and rice (Vishnu-Mittre *et al.* 1972).

At Paunar in Maharashtra during the early and late Satavannah period *Sorghum* was more abundantly used than rice but from the third to the eighth centuries A.D. during the Vakataka–Vishnukundin times the use of rice became so abundant that *Sorghum* became rare (Vishnu-Mittre and Gupta, 1968).

These local changes reveal how cultural contacts could introduce new elements in the food economy of the people.

Note added in proof
On further evidence it appears that the tentative identification of *Eleusine indica* may have to be revised.

2
Crops of west Asia

Wheat

M.V. RAO
I.A.R.I., New Delhi

Introduction

Wheat is one of the foundation crops of Indian agriculture. Carbonised wheat grains obtained from the excavations at Harappa and Mohenjo-daro (2300 B.C.) indicate that this crop was a staple food in the Harappan civilisation. Since the genus *Triticum* has its origin in west Asia it may be concluded that it was one of the crops which spread to the Indian subcontinent in the very early stages of development of organised agriculture. Wheat is now the second most important cereal in India, and the crop plays a very important role in the economy. As early as 1912 Leake and Prasad said, 'The wheat crop of north India is one of the utmost importance to the country, and any improvement either in the value of produce or in the average outturn cannot but lead to direct benefit of no small degree.'

Wheat is essentially a winter crop grown and harvested during the period from October to May. Small areas are sown in the summer months in the hills of both south and north India but production from this summer crop is very small. Annually about 16–17 million hectares are sown, which is about 15 per cent of the area devoted to cereals. About 50 per cent of the area receives irrigation. No country in the world has as much wheat land under irrigation as India. The remaining half depends upon water stored in the soil and upon natural rainfall, which is often scanty or erratic.

The recent wheat revolution is a landmark in the agricultural economy of India. Past restrictions on the distribution of the crop have been greatly relieved. It is now grown in all the states of India. With the availability of photoinsensitive and early-maturing strains it has become one of the most important constituents of multiple cropping systems. The average yield per hectare, which was only seven quintals in 1956, has risen steadily, and has been about 13 quintals in recent years. Where an assured water supply is available yields are high, some progressive farmers reaping as much as 80–5 quintals per hectare with the recently released dwarf varieties. From 12.4 million tonnes in 1964–5, which was the most favourable year before the high yielding dwarfs were taken up on a large scale, the production in the four years 1967–8 to 1971–2 was successively 16.5, 18.0, 20.0 and 23.1 million tonnes. Indeed, wheat production has increased four-fold in India during the last twenty-four years. This high and rising production has greatly helped in stabilising the economy of the country.

Wheat species grown in India

Triticum aestivum L. emend. Thell. the common bread wheat, *T. durum* Desf. the macaroni wheat and *T. dicoccum* Schubl, the emmer wheat, are the three important species grown in India. *T. sphaerococcum* Perc. which was cultivated in the past in Punjab, Uttar Pradesh and Madhya Pradesh has now gone out of cultivation and has been replaced by the more high-yielding bread and *durum* wheats. *T. compactum* was reported to have been cultivated in India (Howard, 1916) but subsequently this was identified as *T. sphaerococcum* (Percival, 1921). Howard (1916) reported the presence of *T. turgidum* L., the rivet or cone wheat, in the samples collected in Baluchistan (now in Pakistan).

The common bread wheat is by far the most important in India, accounting for 86 per cent of the total wheat area. It is raised throughout India. Its concentration is greatest in the northern and the central states of India. In recent years its cultivation has been extended into traditionally rice-growing areas in West Bengal, Andhra Pradesh, Orissa and Tamil Nadu. *T. durum,* which occupies 13 per cent of the wheat area, is grown mostly in central and peninsular India. Some *T. durum* is grown in West Bengal but it is gradually being replaced by *T. aestivum.* At one time the northwestern part of India (now partly in Pakistan) grew *T. durum* but it is no longer under cultivation in Punjab and Haryana. *T. dicoccum* accounts for less than 1 per cent of the wheat area, and is restricted to the states of Andhra Pradesh, Tamil Nadu, Mysore, Maharashtra and Gujarat (Anon. 1963).

The origin of wheat and its evolution in India

Wheat was already a cultivated plant before recorded history, and carbonised wheat grains have been found by archaeologists in the earliest agricultural settlements of Neolithic times. Seeds resembling those of *T. boeticum, T. dicoccoides* and *T. dicoccum* were obtained from the ancient site of Jarmo in Iraq (6700 B.C.) by Robert Braidwood of the University of Chicago, U.S.A. Ancient caves in Europe and mummies in Egypt yielded wheat grains which were considered to be of a later period than those found at Jarmo. Extensive references were made to this cereal in ancient Indian scriptures. Atharva Veda which is supposed to have been written between 1500–500 B.C. (Raychowdhury, 1964) refers to this grain. Through the writings of Theophrastus (325 B.C.) and Columella (first century A.D.) we can surmise that wheat had been cultivated long enough for many of the plant's chief characters to be matters of common knowledge (Clark, 1936). In the absence of any recorded history we cannot now exactly say how, when and where each of the wheat species has originated. Considerable light, however, has been thrown on these aspects through botanical and archaeological studies, and also through expeditions to the centres of origin of the genus.

Triticum durum is a tetraploid carrying genomes designated A and B. *T. aestivum* and *T. sphaerococcum* are hexaploids carrying the A and B genomes, with genome D added. Cytogenetic studies have shown that the A genome was contributed by *T. boeticum,* B by *Aegilops speltoides* and D by *A. squarrosa.* Through chromosome doubling, mutation, intercrossing, and natural and human selection, the species and varieties of wheat have arisen. The home of the parent species is in western Asia, and in western and central Asia the cultivated forms have arisen.

Pal (1966) reported that the northwestern sector of the Indian subcontinent, between the Himalayan range and the Hindukush mountains, is the original home of the common bread wheat *T. aestivum.* Earlier Russian expeditions and more recent expeditions by Japanese and British workers indicated a great amount of diversity in the species in this region. Though recent archaeological work has shown that bread wheat was cultivated in west Asia long before the date of the earliest finds in India, (Helbaek, 1970), it remains certain that the northwest of the Indian subcontinent is a major centre of diversity of the species. *T. sphaerococcum,* which is a wheat of great antiquity and which has been found in the excavations at Mohenjo-daro (now in Pakistan) dating back to 2300 B.C., is supposed to have originated in the northwestern area of the Indian subcontinent. It appears that in ancient India cultivation of *T. sphaerococcum* was quite widespread. Grain samples of this species were also obtained from excavations of ancient sites in Madhya Pradesh and Maharashtra. Singh (1946) opined that due to its high resistance to drought *T. sphaerococcum* was particularly selected by our ancestors. *T. sphaerococcum* appears to be a derivative of *T. aestivum.* According to Ellerton (1939) it arose through the deletion of a small block of genes of *T. aestivum.* Other workers have reported that it originated through a mutation in chromosome 2D of *T. aestivum.* Schmidt and Johnson, and Swaminathan have stated that the *sphaerococcum* characters are governed by a semi-dominant and a recessive gene respectively. Different dosage effects of the so-called 'S' locus were also reported (Swaminathan, 1963).

Howard (1916) mentions a *durum* wheat, *Mecca Muzzama ghanam* (Mecca wheat) which was found in Baluchistan, Khorasan and the Kurram valley. This wheat was apparently introduced into India by Hajis returning from Mecca. It was considered to be sacred and was sown after prayers. Similarly, *T. dicoccum* may have entered by the west coast of India through traders from west Asia. Its cultivation in the states of peninsular India and Gujarat and its absence in the main wheat areas of northwestern, northeastern and central India supports this contention. It is not known how *T. durum* has come to be cultivated in West Bengal since it is not grown in the adjoining areas of Bihar and eastern U.P.

Chromosomal changes have influenced to some extent the differentation of the Indian wheats. For example, Bhaduri and Ghosh (1954*a*) reported the occurrence of two chromosomal biotypes in *Khapli* emmer (IC 834). Biotype I had two pairs of medium-sized, submedianly constricted chromosomes with secondary constrictions, while Biotype II had only one pair. Certain morphological characters such as plant height, ear length, pollen size, pollen sterility and grain setting were found associated with the chromosomal differences of the two biotypes. Bhaduri and Natarajan (1956) found that these differences between Biotype I and Biotype II segregated in Mendelian fashion and that reciprocal differences existed in seed set when Biotype I or Biotype II were used as male or female parents. The one time popular bread wheat variety Pb. C. 591 was also considered to be a chromosomal mutant (Bhaduri and Natarajan, 1956). These spontaneous structural changes or mutations have also been reported in WC 614, a hexaploid local wheat of U.P. Patil and Deodikar (1968) reported differences in differential condensation of chromosomes, formation of bridges, laggards, and micronuclei in crosses involving *T. durum* and *T. dicoccum.* The differences were greater when *T. dicoccum* was the female parent. No differences however, in fertility coefficient in terms of seed/ovule ratio were noticed. Banerjee and Swaminathan (1964) reported that certain mutant loci affected the anatomical characters of the plant as well. One of the old commercial wheats in India, NP 111, was reported to have originated as a mutant.

Natural crossing in wheat, which is very rare in humid climates, is reported to be high in dry regions (Howard and Howard, 1909a; Howard, Howard and Khan, 1910). In the dry tracts of northwestern India natural cross pollination was found to be up to 3 or 4 per cent. This must have contributed greatly to gene exchange and reassortment of variability in Indian wheats. Some of the local types produce abundant pollen and thus may have contributed to natural cross pollination. After systematic wheat improvement work started in India towards the beginning of this century a number of varieties which arose through natural crossing have been reported. Some of these are NP 114, NP 775 (natural crosses on Federation and NP 4) and Vijay (natural cross between Motia *durum* and *Khapli dicoccum*). Howard found a baffling array of types in cultivators' fields of Baluchistan and other areas and she found a large number of intermediate forms of *T. aestivum, T. sphaerococcum, T. durum,* etc. To quote Howard, 'In the case of *T. compactum* [apparently *T. sphaerococcum*] and *T. vulgare*, the distinction based on the glume and grain shape breaks down utterly. In the Himalayan tracts, a large number of wheats are found which combine the ear shape and the general characters of *T. vulgare* with glumes and grains as rounded as in a typical *compactum.*' Howard and Howard (1909b) postulated that the famous high quality Punjab Local Type 9 wheat, is a cross between *T. durum* and *T. sphaerococcum,* since it combined the characters of both these species. In view of the difference between them in chromosome complements, gene exchange would be restricted, but the Howards' description indi-. cates the extent and significance of character recombination in the phylogeny of the Indian wheats.

The different species grown in India have adapted to different ecological conditions. Thus *T. sphaerococcum,* because of its drought resistance, was grown in areas of northwestern India under inundation moisture and little rain. *T. durum,* all Indian races of which are susceptible to yellow rust (*Puccinia striiformis* West.), has become confined to areas of central and peninsular India where yellow rust is not a problem. *T. dicoccum,* because of its high resistance to a wide range of rusts, survived in peninsular India where other wheat species could not. *T. aestivum* is grown in north India because of its responsiveness to intensive management practices, in an area where assured irrigation water and high intrinsic soil fertility makes high productivity possible.

The environment of the Indian wheat crop

The dominant environmental conditions that influence the growth of the wheat crop in India are the maximum duration of the growth period in each tract, and the available soil moisture during that growth period. The effective growth period for wheat is terminated by the onset of hot weather. This varies with the latitude. In peninsular India wheat ripens by the middle of February, in central India by the end of February or beginning of March, in Uttar Pradesh in March–April, in Punjab in May and in the northern hills in June. Moisture greatly influences the length and strength of straw, yield, quality of grain and to some extent the ripening. This overall pattern is modified by seasonal variations such as late rains, early onset of hot winds, frost, and heavy dews. Within this climatic pattern, which gives a growing season varying from four to eight months according to locality, Indian wheats have generated a range of varieties that match the local circumstances of the tracts in which they are grown. Compared with the spring-sown varieties of temperate regions they offer a group of short-duration types that has been of great value in breeding for earliness in other countries beside India.

Besides adaptation to the major climatic factors that determine the length and nature of the growing season, Indian wheats have been subject to selection, both natural and human, to improve crop performance. The most important of these selective forces are (1) tolerance of adverse soil conditions, (2) disease resistance, and (3) improved harvesting and quality characteristics.

Tolerance of adverse soil conditions

For many centuries Indian wheats have been grown at low fertility levels and since there is little precipitation in the wheat growing season, they have often suffered moisture stress. Many local wheats are drought-tolerant. They formed the base for further improvement work in breeding wheats for rainfed areas in India. Some of them have a local adaptation for high and non-synchronous tillering, spreading habit in the earlier stages of growth, and a short ripening period (from heading to ripening) to suit the rising temperature conditions of the spring season. Large tracts in India, particularly in northwestern India, suffer from high soil salinity. Some wheats have been evolved to meet these conditions. One such wheat with a high tolerance of saline conditions is Kharchia in Rajasthan. Since this wheat is susceptible to rusts, the wheat breeders in Rajasthan developed the resistant Kharchia 65 through back-cross breeding with a resistant type.

Disease resistance

The most important wheat diseases in India are the rusts. All three wheat rusts, namely, the black or stem rust (*Puccinia graminis tritici* (Pers.) Erikss. and Henn.), brown or leaf rust (*P. recondita* Rob. ex. Desm.) and yellow or stripe rust (*P. striiformis* West.) are met with in India. Black and brown rusts are spread all over the country while yellow rust is more restricted to the cooler regions of the north. These rusts have often appeared in epidemic proportions in different parts of the country in the past, and caused widespread misery and suffering to the farmers. Although the local Indian wheats, perhaps with the exception of the *Khapli* emmer, do not have a high degree of rust resistance, some of them do have a measure of tolerance or field resistance to the rust races present in a given area. Thus, Howard, Howard and Khan (1922) found in 122 local samples from Bihar and Orissa, certain types having a good measure of field resistance to brown, yellow and black rusts. Similarly in Punjab, where yellow rust is very important, the local wheats had a greater measure of tolerance to this rust. In the emphasis in recent breeding work on resistance controlled by hypersensitive genes, we may be losing these genes for tolerance which have been built into the local wheats over a long period of evolution.

Another disease which has of late become very important in India is Alternaria leaf blight caused by *Alternaria triticina*. This disease was first noticed with the release of improved wheats carrying foreign gene material. When Kenya wheats were used to incorporate black rust resistance into local wheats of Maharashtra and Gujarat, Alternaria blight suddenly became a disease of major importance. It appears that the disease must have long been present in India, so long indeed that the local strains have built up high resistance to it.

Improved harvesting and quality characteristics

Indian wheats generally are more resistant to shattering than their counterparts in other parts of the world. The wheats in northwestern, central and peninsular India are more resistant to shattering than those grown in the more humid areas of eastern India or in the northern hills. Awned wheats were preferred in the past over the awnless wheats because they were considered more resistant to shattering (all local awnless wheats were invariably susceptible to shattering – Howard and Howard, 1909*b*). Because of the non-shattering character, the harvesting of the Indian wheats could be delayed slightly (labour is dear and scarce at harvesting time). Awned wheats are also considered to be less liable to bird damage, although this is not always true. In the hills of north India, because of the high humidity that prevails at harvest time, awnless wheats, which thresh easily, are preferred to the awned wheats. Chaff characters also vary with local climate, pubescent or felted chaff being unsuitable in damp climates. Pubescent or felted wheats were more common in the climate of Punjab than in the damper wheat-growing tracts of U.P. In Bengal, with a high humidity, the felted wheats are very rare.

Quality is one of the factors that influenced the cultivation of specific types of wheat in different parts of India. It is affected by climate, soil and water management, as well as by variety, and as a result certain areas have gained a reputation for particular qualities in the wheat they produce. A host of local Indian names indicate either the quality, the place of cultivation or the uses to which a particular type of wheat is put. Farmers in the past grew both red and white wheats. Many enquiries in the villages in the Indo-Gangetic plain have elicited the information that for his own use the cultivator preferred hard wheats often hard reds (Howard and Howard, 1909*b*). However, due to trade considerations, the emphasis gradually shifted to white wheats in the last century. Colour considerations became so ingrained in the Indian trade that the hard amber wheat always fetched a premium in the market over the red hard, red soft or white soft wheats.

Indian wheat, as it was grown in the past before wheat improvement work, used to be a mixture of several 'sorts' or types. Pure line selection led to the isolation of a number of high-quality wheats that gained an international reputation. They formed a basis for further work, both in India and in other parts of the world. Farrer, in Australia in 1899, was one of the first to recognise the high quality of some of the Indian wheats, and he used them extensively for the improvement of Australian wheats in quality and in earliness. A Bihar local wheat which was one of the constituents of Hard Red Calcutta went into the production of high-quality Marquis wheat (Howard *et al.* 1922).

This brief survey of the diversity of types that has arisen under the selection pressures of the Indian environment emphasises the great value of the germ plasm of the indigenous Indian wheats. There is an urgent need to collect, evaluate and preserve this material, before it is lost under the impact of the modern high-yielding varieties.

The influence of man on the development of Indian wheat

With the establishment towards the beginning of this century of the Indian Agricultural Research Institute at Pusa, Bihar, and the Agricultural Research Institutions at Lyallpur and Cawnpore, systematic improvement work began in India. Most of the wheats grown today have been developed through systematic breeding. What has been achieved during the last sixty-five years has been comprehensively dealt with from time to time (Howard and Howard, 1909b; Murty 1958; Pal, 1966; Kohli, 1969; Anon. 1968). The work done during this century may be summarised as follows:

First there was a period during which the local mixtures were sorted out and comprehensive series of pure lines were established. The collections that were made were extensive, and represented the stocks in cultivation in most of the wheat growing areas. Some of these pure lines yielded commercially valuable types that have been widely grown. Many of them have excellent grain and quality characters. Following this early work Indian wheats, which had commanded a low price on the world market, gained a better reputation for quality.

There followed a period during which hybridisation was used extensively to further improve yield and the quality of the grain. Systematic hybridisation work was initiated both at intra- and interspecific level. The high yields obtained from winter wheats in Britain and other European countries tempted Indian wheat workers to use these wheats either for direct introduction or for hybridisation, but without success. Experience has shown that better and quicker results can be obtained by intercrossing pure lines isolated in India or by using wheats of similar maturity periods such as those from Australia (some of which have Indian germ plasm in their ancestry). A number of successful wheats were developed by crossing local × local or local × foreign spring wheats and these held the field in India for a long time. However, these wheats, in spite of their other acceptable attributes, were susceptible to disease, particularly the rusts, and had to be replaced by disease-resistant varieties.

Specific mention may be made of the role of interspecific hybridisation in the wheat improvement work in India. Some species were deliberately crossed with species having different chromosome numbers with the hope of getting derivatives that combined good characters of both. In some cases three species were crossed (e.g. Niphad 4 which was at one time prized for its grain type in Maharashtra was a derivative of a cross involving *T. durum, T. dicoccum* and *T. aestivum*).

Several Indian wheat breeders used *T. durum* and *T. dicoccum* to improve the *aestivum* wheats. The primary objective in using the *durums* was to incorporate their drought-resisting character, bold grain and to some extent resistance to black rust. The hybrid populations were systematically studied and experience indicated that, although a wide range of chromosome numbers was observed, individuals with euploid numbers were stable. When large segregating populations were grown, combinations which had the desirable characters of the two species involved in a particular cross were picked up. A number of successful varieties were produced in Madhya Pradesh and Maharashtra.

Table 1   *Diseases attacking the wheat crop*

Disease	Causative organism	Region of India where most severe
Black rust	*Puccinia graminis tritici*	All
Brown rust	*P. recondita*	All
Yellow rust	*P. striiformis*	North (hills and plains)
Loose smut	*Ustilago nuda*	North
Bunt	*Tilletia foetida* and *T. caries*	Hills in the north
Karnal bunt	*Neovassia indica*	Northwest
Flag smut	*Urocystis tritici*	Northwest
Powdery mildew	*Erisyphe graminis*	Hills, north and south
Leaf blight	*Alternaria triticina*	East and southwest
Leaf spot	*Helminthosporium sativum*	Eastern
Leaf blotch	*Septoria tritici*	Southern hills
Glume blotch	*S. nodorum*	Southern hills
Tundu disease	Combination of *Corynebacterium, tritici* and *Anguina tritici* (nematode)	Northwest
Ear cockle	*Anguina tritici*	Northwest
Root rot	*Rhizoctonia*	Central
Foot rots	Species of *Helminthosporium, Fusarium* and *Pythium*	East

The performance of some of the *T. durum* × *T. aestivum* derivatives under deficit moisture conditions clearly indicates that the *durums* do transmit drought-resisting properties. All the *aestivum* wheats which are known to perform well under low moisture conditions have *durum* wheats in their ancestry. In the present emphasis on increasing yields under non-irrigated conditions a fresh approach involving crosses between *T. aestivum* and *T. durum* would be rewarding.

Wheat in India is ravaged by a number of diseases, the more important of which are listed in Table 1. There is a definite geographic distribution of these diseases. For example, the bunts are prevalent in the hills of the north, while *Septoria* is restricted to the southern hills. Flag smut, Karnal bunt, Tundu, and ear cockle diseases are found in the northwest while *Helminthosporium* leaf spots and foot rots are restricted to eastern India. Loose smut is predominant in the north. Black and brown rusts as mentioned elsewhere are found all over India, while yellow rust is restricted to the hills and plains of the north.

Diseases, particularly the rusts and the smuts, received considerable attention from the mid-thirties onwards, both from plant pathologists and from plant breeders. The work of the late Dr K.C. Mehta resulted in the isolation of physiologic races of the three rusts, and also in explaining how rust attack recurs every year in the plains of India. It was early recognised that the prospect of avoiding losses from disease lay in breeding resistant varieties. In the absence of genetic stocks resistant to all the three rusts and the smuts, particularly loose smut, the work at the Indian Agricultural Research Institute was directed to the objective of breeding for resistance to all the three rusts and loose smut, in successive steps. The wheats extensively used in breeding for disease resistance came from Australia, Kenya, the U.S.A., Italy and Latin America, together with the internationally assembled rust differentials.

T. dicoccum has also been extensively used in Indian wheat improvement work.

It is known for its high resistance to several diseases, particularly the rusts, powdery mildew, etc. *Khapli* wheat, which is valued in many countries for its high resistance to black rust and powdery mildew, is a local *dicoccum* wheat of Maharashtra. In India, selection in local material yielded wheats like NP 200 and NP 201 which have high field resistance to brown rust, besides being resistant to black rust and powdery mildew.

In recent years *T. carthlicum, T. pyramidale, T. polonicum* and *T. timopheevi* have been used by Indian wheat workers primarily to incorporate disease resistance to *durum* wheats. The derivatives of these interspecies crosses are now in yield performance tests at the All-India level. The I.A.R.I. work resulted in the development of the NP wheats of the '700 and 800 series'. Similar projects in other wheat-growing states also resulted in the release of new resistant varieties. Many of these wheats were superior in quality, and resistant or tolerant to disease. They were also deliberately bred with tall straw, to meet the need of the huge cattle population for forage. At that time fertilisers were not available in sufficient quantity and intensive agricultural practices were not followed. In those circumstances the advantage of long straw in forage production was not offset by the tendency to lodge under high fertility conditions. Even so, some of the earlier wheats stood well enough to give yields up to 50–55 quintals per hectare under favourable conditions.

The vegetative duration of many wheats, including those indigenous in India, is governed by a physiological response to day length. The great advance in exploiting high fertility conditions came from the availability of stocks that were both insensitive to day length and short strawed to resist lodging. The story of the introduction and spread of the photoinsensitive Mexican dwarf wheats carrying the 'Norin' genes for short straw has recently been reviewed (Anon. 1968). Dwarf mutants have long been known in wheat. Indeed, the most ancient wheat of India, *T. sphaerococcum,* carries a dwarfing gene. However, this and most other dwarfing genes produce also undesirably dense and compact ears, and have been of no value in breeding productive varieties. The dwarfing genes in the Norin stock, on the other hand, do not adversely affect the size and form of the ear.

In 1962 a study was carried out at the I.A.R.I. of the results of recent breeding work and of the constraints on the evolution of highly productive varieties under Indian conditions. The study showed up clearly three major limiting factors:

1. The morphology and developmental rhythm of the long-strawed varieties then in cultivation were not suitable for the better fertility levels that were readily established by the use of fertilisers.
2. Varieties with tall straw and a proneness to lodging went down badly when irrigated in the latter part of the season.
3. Only the more recent products of breeding schemes were resistant to rusts and to loose smut.

The significance of lodging in delaying ripening and impeding grain development was apparent not only at high fertility levels, but also as a consequence of late season storm damage at all fertility levels, and this led those concerned with breeding policy to look closely at the possibility of introducing short-strawed, lodging-resistant material. At that time stocks carrying dwarfing genes derived from the Japanese Norin wheat were becoming available. In these stocks stem length was reduced and re-sistance to lodging was high, while at the same time there were no unwanted side effects on the ear. Norin wheat was introduced into the United States and dwarf winter wheats were bred from crosses with American varieties. Dwarf spring wheats were bred in Mexico from Norin material at the Rockefeller station at about the same time. In 1962 a request was made to the Rockefeller Foundation for seed of these stocks for trial in India. In 1963 Dr Borlaug visited India and arranged for seed of four dwarf stocks to be sent for trial. The spectacular yields of these stocks led to arrangements for seed multiplication in the summer season in the Nilgiri Hills, and widespread testing in the following *rabi* (winter) season. Two varieties, Lerma Rojo and Sonora 64, were released for general cultivation in 1965, and the agronomic practices necessary for successful cultivation were standardised and publicised. Success was so great that the Indian Government imported 18 000 tonnes of seed in 1966. From then on the spread of the Mexican dwarfs in the irrigated wheat belt went on with extraordinary speed.

The dwarf stocks that were introduced were hybrids of Norin dwarfs with Mexican spring wheats. They had been bred in Mexico with alternate generations in different climatic and day-length regimes, primarily in order to get two generations a year. A valuable side effect of this system was to establish a good degree of insensitiveness to photoperiod. Selection for disease resistance had also been prac-tised, and the stocks introduced were found to show a remarkable level of resistance under Indian conditions. A further important feature of the original stocks was their diversity. They had not been bred to pure line standards, and there remained in them a reservoir of genetic potential that Indian wheat breeders were quick to exploit. Varieties differing in duration were selected to meet the needs of differing climate and cropping circumstances. Rust-resistant stocks were selected from seg-regating material. Grain types were picked out more nearly approaching those pre-ferred in the Indian market.

The most successful material was derived from a population designated S 227 from a Mexican cross Penjamo sib × Gabo 55. Selections made from this material at the I.A.R.I., P.A.U., Ludhiana, and U.P.A.U., Pantnagar made up the stock that was distributed under the name Kalyansona and that in a few years became the dominant variety in the irrigated wheat areas. It is of interest to note that selections from the same cross have been highly successful in Mexico and Pakistan.

This period has been followed by a broadening of the breeding system, and the emergence of a new and wider range of varieties in the next generation. The primary stimulus for a new breeding advance was the fear that the rust resistance of the current varieties might break down. With the dominance of Kalyansona, a breakdown in that one variety might lead to a major disaster to Indian food supply. Other considerations were the possibility that an advance might come from the introduction of more dwarfing genes from the Norin stock, and the need to establish a grain quality up to the standards for which the Indian market will pay the best price. This programme has involved very extensive hybridisation between Mexican dwarfs and indigenous Indian varieties. The Norin dwarfs were believed to carry three dwarfing genes, having cumulative effects. It has been shown that more than three chromosomes are involved in reductions in height, but three levels of dwarfing can be roughly distinguished morphologically. These are known as 'single', 'double' and 'triple' dwarfs. The first generation of dwarfs comprised single and double dwarfs. Kalyansona is a double dwarf. New combinations of Mexican and Indian germ plasm are now becoming available, including triple dwarfs.

The impact of the dwarf wheats has hitherto been almost entirely on the irrigated wheat lands. They have made it possible to exploit high fertility levels established by the use of fertilisers, and favourable water regimes by timely watering right up to crop ripening. On the rainfed wheat lands, the water regime is not under control, and fertiliser use is restricted by the risk of failure due to water shortage late in the season. The risks attending high production on rainfed lands can be reduced by breeding short-term varieties, and improvement of disease resistance. Short straw and resistance to lodging are a prerequisite for any advance in productivity. Thus the Norin dwarf stocks have contributed to the current breeding work for unirrigated areas. Dwarfing genes and rust resistance genes have been transferred from the hexaploid Mexican dwarfs to the tetraploid Indian durums. Results from current All-India co-ordinated trials show that the new dwarf *durum* material has a high yield potential, equalling or even exceeding that of similar *aestivum* wheats.

The magnitude of the break-through associated with the dwarf wheats is indicated by the increase in production. Over the seven years 1961–7, production averaged 11.1 million tonnes per annum. It then increased rapidly, and in 1971 was 23.2 million tonnes, or rather more than double. To the plant breeder, the nature of the genetic resources on which the evolution of these new wheats was based, is of great interest. The Norin dwarfs were Japanese wheats. Both long-term winter wheats and short-term spring wheats from American stocks were crossed with them. The resultant Mexican stocks have now been hybridised with advanced Indian varieties. Thus the diversity available in a number of the advanced wheat-growing areas of the world has been drawn upon in the synthesis of a new wheat stock, designed morphologically and physiologically to exploit high fertility conditions. This emphasises the importance of collecting and maintaining the land races of a country such as India. The sweeping success of a variety like Kalyansona can put in jeopardy the genetic material on which further success must be built. At the present time the preservation of this material is a high priority for Indian plant breeding.

The future prospects for wheat breeding
The increasing demand for food grains in India will put continuing pressure on our attempts to increase wheat production. Since the land available for raising food grains is limited future prospects revolve round two alternatives: increasing production per unit area whether irrigated or unirrigated, and controlling disease and pests that limit production. Quality aspects, particularly the nutritional, will receive increasing attention.

Although the average yield per hectare has increased from 7 to 13 quintals in the last few years, still this figure is low as compared to the yields obtained in advanced countries such as Britain (41.3 quintals per hectare), West Germany (37.9), France (34.4), Mexico (28.4), Canada (20.9), etc. There is ample evidence from national demonstrations and farmers' competitions that with the right inputs of fertiliser and irrigation, much higher yields are possible with the current dwarf varieties.

These wheats, moreover, being photoinsensitive and early maturing, fit in well to the multiple cropping pattern that is increasingly adopted in Indian irrigated agriculture. Thus on the irrigated land the task of the future is to consolidate and exploit the gains already made.

Increased disease resistance remains a high priority in the breeding programme. The food situation in India is so narrowly balanced that the country cannot afford the risk of epidemic disease. Much is now known of the foci of persistence of diseases, and of the nature and circumstances of their spread. It is now possible, for example, to concentrate on breeding wheats with broad-based resistance to black and brown rusts for the Nilgiri and Palni hills in south India where these two have their over-season survival, and to yellow and brown rusts for the northern hills where they also survive from one season to the next. Similarly, as locally adapted varieties are bred, resistance to the locally prevalent diseases will be incorporated, and some saving in breeding effort achieved by disregarding diseases that do not present a hazard in the region for which the variety is bred.

The challenge of the rainfed crop must now be taken up, in terms of agronomy and pest and disease control as well as of breeding. The existing better yielding types are characterised by non-synchronous tillering, medium maturity, reduced period for grain ripening (once the flowering has taken place) and relatively bolder grain. The yielding ability of these wheats may be further improved by utilisation of interspecific hybridisation (particularly with the *durums*) and the winter wheats from temperate climates, with their better root system and winter hardiness (which is another form of drought resistance). Their use in Indian wheat breeding will not only bring in diverse genes for yield, but will enable the breeders to select for needed maturity periods.

The *durum* wheats, which are so far restricted to the unirrigated areas of central and peninsular India, will be taken up for cultivation in northern India also where they are conspicuously absent at present. This is now possible for two reasons:

1. Indian *durums* have been unsuccessful in north India as they are susceptible to yellow rust. Dwarf *durums* that are resistant to yellow rust are now available.
2. These new dwarf *durums* appear to have a higher yield potential than the dwarf *aestivums.*

Wheat has already spread to some of the traditional rice-growing areas. In the winter season, provided it is rust resistant, it is more reliable than rice or any other alternative crop, and it is likely to spread further. Early maturing varieties are required, to fit in to a rice/wheat rotation.

Quality considerations will be of great importance. Because of the limited availability of products of animal origin, most of the nutritional needs of the Indian population are met by the cereal crops such as rice, wheat, maize and millets. The future wheats will be nutritionally better in terms of protein, and of some of the essential amino acids. Varieties will be bred to meet the needs of the fast growing milling and bread-making industries, as well as for the traditional unleavened pan-baked *chapati,* which will remain the main form in which wheat will be consumed in India.

The great diversity of uses to which wheat is put can be met by varieties of the two chief wheat species, *T. aestivum* and *T. durum*. With the improvement of the productivity of Indian farmlands there will remain no place for primitive low-yielding types, and it is to be expected that *T. dicoccum* will go out of cultivation, as *T. sphaerococcum* has already done.

Barley

J.S. BAKSHI and R.S. RANA
Division of Genetics, I.A.R.I., New Delhi

Archaeological evidence shows that at the beginning of agriculture, barley was the most important crop grown, exceeding in acreage and value the other cereals. The cultivation of barley at present extends from subarctic to subtropical regions and barley is Man's most dependable temperate cereal crop where alkali soils or drought conditions are encountered.

Origin and antiquity

All barleys, wild and cultivated, belong to the same potentially interfertile population and are grouped under one species, *Hordeum vulgare* L. emend. Bowden. There is now overwhelming evidence to show that barley was domesticated from a two-row progenitor resembling *H. vulgare* sub sp. *spontaneum* in the Near East region around 8000 B.C. (Harlan, 1968, 1969). Barley culture in India appears to have come from western Asia (the Near East in European terms), and can be followed with fair certainty across northern India and then southward (Raikes and Dyson, 1961; Sankalia, Subharao and Deo, 1953; see also Vishnu-Mittre, above). Bread wheat as well as barley grains have been found in excavations of the Indus civilisation site at Mohenjo-daro (*c.*2300 B.C.), and both were cultivated forms. The series of archaeological finds from Ur and Mohenjo-daro have now pushed back the antiquity of India's relations with western Asia to the third millennium B.C. At that time the ancient Indus cities were in regular and intimate contact with the Sumerian cities of Iraq (Sen, 1963).

The Harappan culture seems to end abruptly with the arrival of the Aryans in about 1500 B.C. Barley, called *Yava* in Sanskrit, is the most frequently mentioned cereal in the earliest records of the Indo-Aryans and was probably the principal staple food during the Vedic period (1500–600 B.C.), as can be seen from numerous references in ancient Aryan literature, namely the *Rig Veda,* the *Atharva Veda,* the *Yajur Veda* and also in the *Upanishads.* Wheat finds definite mention in the *Atharva Veda* which was composed considerably later than the *Rig Veda.* In fact, the composition of the *Atharva Veda* is considered to reflect the permeation of Harappan knowledge into that of the Aryans. Thus, although the Indo-Aryans appear to have brought with them their staple food grain (*Yava*), yet the wheat and barley material of the Indus people was also adopted by them, generating new variability required for more intensive cultivation. It may well be that barley came to India not once but on several prehistoric occasions, either through the trade routes or along with immigrating people.

The place of barley in Indian agriculture

Barley at present is grown widely in India. In acreage it comes next to wheat as a winter cereal crop and occupies approximately 2.4 million hectares, producing nearly 2.8 million tonnes of grain annually. In areas where winter rainfall is deficient or where soils are alkaline, barley is grown in preference to wheat. The chief barley-growing region is northwest India, with substantial areas grown as far east as Bihar and as far south as Madhya Pradesh. In plains and at low altitudes in the hills only six-row hulled barleys are grown commercially. At higher altitudes, however, where barley replaces wheat as a staple crop, hull-less six-row types are grown.

The bulk of the barley grain in India is used as human food. It is consumed either as flour (mixed with wheat) for *chapati* or as parched grain in the form of *sattu*. A small portion of the produce is utilised in industry for malting and brewing. If a farmer has some surplus grain, he uses it for livestock feeding.

Throughout the barley-growing area, wheat is the preferred winter cereal, and barley is confined to areas too dry or too saline to carry a satisfactory wheat crop. Hence on the one hand the chief incentive in crop improvement has been to work on the wheat crop, and on the other, recent improvements in water use and in fertiliser practice have tended to bring in wheat to the better barley lands and to leave barley still confined to poor lands, and hence a low-yielding crop.

These are the main reasons why there has not been the outstanding improvement in performance in the barley crop that has been achieved with wheat. Indeed improvement work on barley in India has been small in volume, and only begun relatively recently. Initially, attempts were directed to the isolation of pure lines in the mixture of different morphological and agronomical types grown locally. Selection was mainly practised for higher yielding capacity, better head and grain types, earliness, tolerance to diseases and local adaptation (Bose, 1931). As a result of these efforts, a series of new varieties was developed in Uttar Pradesh, in Punjab, in Bihar and Rajasthan and at the I.A.R.I., Pusa. The next phase of improvement was intervarietal hybridisation mainly between the existing indigenous varieties. It resulted in the issue of a second series of new varieties. These varieties, though superior in yield to local mixtures or local sorts, were characterised by a narrow range of adaptation, and susceptibility to diseases and lodging. Some are, however, reported to have grain of acceptable malting quality.

Recent barley breeding work

The barley breeding work described above was carried out with limited objectives in view and the varieties released lacked many desirable features. They had no significant impact on yield per acre, which remained more or less stationary up to about 1965.

For current breeding work the barley-growing areas in the country are grouped into four broad regions, the northern hills, the northwestern plains, the northeastern plains, and the central plains. These zones represent the chief agroclimatic conditions under which barley is grown. In the co-ordinated breeding programme now in operation, environmental stability in the three plain zones has been studied, and a system of yield trials established whereby the products of breeding work are tested throughout the barley-growing area. The performance of varieties currently available has

Table 1.   *Performance of barley varieties under saline conditions*

Variety	Yield/plot under saline conditions (g)	Per cent reduction in yield at 24 mmho cm^{-1} over 8 mmho cm^{-1} (pots)
Ratna	1540	39
RS 6	1506	53
BHD 18	1547	60
NP 113	1096	53
EB 3	1215	43
K 18	1098	56
NP 13	913	55
EC 24882	736	85

been evaluated, and recommendations have been made for rain-fed and irrigated areas in different zones.

Work now in progress falls under the following main heads:

Salt tolerance

Studies of salt tolerance were carried out on a range of varieties, including Ratna (recommended for the northeast plains) and RS 6 (recommended for the central plains), since barley is considered to be the most profitable grain crop where saline or alkaline conditions are encountered. Data from pot experiments show that Ratna is the most salt-tolerant variety (Table 1). This variety showed a reduction of only 39 per cent in yield when salt concentration was raised from 8 mmho cm^{-1} to 24 mmho cm^{-1}.

In contrast, other varieties registered a reduction in yield ranging from 43 per cent to 85 per cent. In the actual field tests under saline conditions, Ratna was among the top yielding varieties (Bains, Singh, Dayanand and Bakshi, 1970).

Disease resistance

The important diseases of barley in India are stripe rust, and *Helminthosporium*. Smuts are significant, but of lesser importance. Aphids are a serious pest. Resistance breeding has received little attention in the past, and many of the available varieties are susceptible to these pathogens. In current programmes genetic resistance is given the highest priority.

Twelve races of stripe rust (*Puccinia striiformis* West.) have been recorded in India. Of these, five races – G, A, 57, 31, and 24 – do noticeable damage under natural conditions. About 3000 barley lines were screened to locate sources of resistance. All Indian barleys were susceptible, and in the whole collection resistance to Indian races was rare. Only five varieties, set out along with their resistance performance in Table 2, offered satisfactory sources of resistance.

Genetical studies conducted with EB 410 showed that field resistance is governed by a single dominant gene while seedling resistance is governed by two genes. EB 145 appears to carry six or seven genes for physiological resistance to the five races, although resistance to individual races is inherited simply (Bakshi and Luthra, 1970). Field

Table 2. *Sources of resistance to important Indian races of stripe rust*

| Genetic stock | | Ear type | Field reaction† | Inheritance of resistance | |
Name*	Origin			Seedling	Field
EB 410	Unknown	Two-row	TS	G, 31, 57, A Two dominant genes	Single gene
EB 145	Unknown	Two-row	TR	G, 31, 24, 57, A 6 or 7 genes	Not studied
EB 1626	West Germany	Two-row	TR-TS	–	–
EB 1556	Ethiopia	Six-row	TR	–	Two complementary recessive genes
Heitpas	U.S.A.	Six-row	TR	–	–

* EB numbers in this and following tables are Indian accession numbers

† Intensity of rust infection and type of pustule (T = Traces; R = Resistant-type pustule; S = Susceptible-type pustule).

resistance in EB 1556 has been found to be controlled by two complimentary recessive genes.

In spite of major genes for resistance and simple inheritance, progress in transferring resistance has been rather slow. One of the main reasons for this is that these sources of resistance have several drawbacks. For example, EB 410 and EB 145 are very late, and have weak straw. Heitpas again is very late. EB 1556 is highly susceptible to aphids. In view of this, there is still a great need for new sources of resistance.

Screening the available germ plasm for response to *Helminthosporium gramineum* attack revealed high resistance in the Indian varieties K 12, K 24, Cn 292, Se 1, 64/24. Among these K 12 is a very good combiner for resistance to this disease. Furthermore, high resistance has also been identified in three introduced varieties, EB 438 of unknown origin, EB 783 from Germany, and EB 928 from Israel.

Pest resistance

About 3000 genetic stocks were tested for resistance to aphids (*Rhopalosiphon maidis* Fitch). Genes for resistance were completely lacking in the Indian material. Of the introduced material tested, one two-row variety from Taiwan, EB 921, gave an immune reaction. Ten accessions, all six-row varieties from the eastern Mediterranean region, were moderately to highly resistant. From the Balkans, EB 2434, EB 2436 and EB 2457 proved to be highly resistant, and EB 841 and EB 2435 resistant to moderately resistant. From Crete, EB 2500, EB 2501, EB 2499 and EB 2508 were resistant to moderately resistant, and from Greece EB 2515 was moderately resistant.

In breeding work EB 921 has mainly been used and it has been possible to transfer its complete immunity easily, indicating that probably resistance is simply inherited. Currently, efforts are under way to combine aphid resistance with yellow rust resistance.

Lodging resistance

Lodging is a serious limiting factor in increasing barley production as most Indian

Table 3 *Dwarf-induced mutants used in the Indian barley breeding programme*

Original stock		Name of mutant	Source	Height (cm)
Name	Country of origin			
NP 13	India	Dwarf 21	I.A.R.I.	83
NP 113	India	BM 10	I.A.R.I.	71
		BM 12	I.A.R.I.	74
Durgapura RS 17	India	RDB 1	Durgapur	89
		RDB 2	Durgapur	93
	Sweden	Erectoid Early 18	Svalof	93

varieties lodge even at low fertility levels. Breeding dwarf or semi-dwarf barleys with good straw strength, responsive to higher doses of fertiliser, is essential to increase barley productivity. Barleys of this kind have been bred extensively in other parts of the world, notably in western Europe and Britain, but advance has come by the accumulation of genes of small individual effect and not from major dwarfing genes as in wheat. There has been a search all over the world for useful dwarfing genes. Attempts, in India and elsewhere, to use the Japanese 'Uzu' dwarfs, have not been fruitful, as the gene has strong pleiotropic effects, reducing in size useful plant parts such as the ear, grain and leaves. Mutations causing dwarfing have been induced both in Sweden and in India. Some of these have shown promise in breeding programmes. Short-strawed varieties carrying induced mutant genes have been released in Sweden and Dr N.E. Borlaug (personal communication) has used in the breeding programme in Mexico the dwarf mutant induced at the I.A.R.I., New Delhi in the variety NP 13. In the Indian breeding programme five dwarf-induced mutants are in use. Information on these is given in Table 3. Among these, two mutants BM 10 and BM 12 appear to be most promising. They are shorter than the others, and have stiffer straw and a better head type.

During 1969—70, about eighty promising varieties from Australia, Canada, U.S.A., Mexico, Sweden, Hungary, Denmark and Japan were introduced. The objective was to locate varieties combining good straw strength with other desirable characters. As a result of a study conducted in 1970 it appeared that varieties from Mexico and spring varieties from the U.S.A. (California) and Canada do well in India and provide useful material. Some induced mutants from Sweden were early, and are useful. Most varieties from Sweden, Denmark, Hungary and Japan were late. Two two-row varieties from Australia were identified as useful for malting. The most promising of the varieties mentioned above are being used in the breeding programme to impart straw stiffness.

Nutritional and malting qualities
Since barley grain mainly goes for human consumption in India, improvement in nutritional qualities is important. Two newly released varieties have over 13 per cent protein as against the range of 9—12 per cent in the earlier varieties issued.

The production of barley grain specifically intended for malting purposes for brewing, distilling, and for patent foods, received scant attention in India until the recent substantial rise in the demand for high-quality malt barley for brewing. This is due to a considerable export potential for high-quality malt and because of the additional breweries now being set up in the country. For malting, grain with high starch and low protein is required, whereas for food barley, a high protein content is superior.

Two-row barleys are preferred for malting, whereas the food barleys of India are all six-row. The introduction of two-row barleys from Australia that respond well to Indian growing conditions, and produce a good malting type of grain, opens up the possibility of breeding two-row barleys for malting while maintaining six-row types for food. Thus it is proposed to use the morphological character of row number to establish a ready market identification of cultivars bred for the widely different biochemical characters required for malting and for human food.

Conclusions

The very limited genetic improvement so far achieved in Indian barleys is in marked contrast to the very great improvements attained in other parts of the world, and particularly in Britain and western Europe. It is clear both from the experience of these countries and from the characters of the range of material now assembled in India, that the constraints on improvement have not been genetic. Barley has been a crop of hard conditions, both of soil and of climate. Wheat, grown under good conditions of soil and of water supply, has been a more attractive crop to the plant breeder, and new responsive varieties have been bred to meet improved conditions of fertility level and water management. And where barley lands have been improved, the tendency has been to replace barley with the new responsive wheat varieties.

It is the aim of the current barley improvement programme to provide new varieties of barley that will give profitable responses to better conditions, and thereby to improve its position in competition with wheat, as well as to meet more successfully the adverse conditions of drought and salinity under which barley has no real competitor. With the greatly improved supply position in the wheat crop, the opportunity for diversification provided by the two distinct outlets for barley — human food and malting — is of great value to Indian agriculture.

3
Crops of south Asia and Africa

Rice

S.V.S. SHASTRY
A.I.C.R.I.P., Hyderabad and

S.D. SHARMA
I.A.R.I. Regional Station, Hyderabad

The genus *Oryza* belongs to the family Gramineae tribe Oryzeae and is represented by plants growing up to 2 m in height. They are mostly hydrophytes and are distributed in the tropics of the world. The genus has about twenty-eight species and subspecies. The species (and subspecies) can be conveniently grouped into three sections, and further divided into seven series (Table 1). The species of different series are morphologically distinct and can be unmistakably identified. On the other hand, the species of the same series are somewhat similar and even expert taxonomists have differed on their identification.

Most species of *Oryza* are diploid, with 2n = 24 chromosomes. Some tetraploid species are known, but not among the cultivated rices and their close relations. Section Padia, which is confined to southeast Asia, and Section Angustifolia, which is confined to Africa, have not contributed to the evolution of the cultivated rices. Section Oryza, which is pantropic in distribution, includes the largest number of species of any section. Section Oryza is made up of two series. Of these, Series Latifoliae with eleven species includes no cultivated types. Series Sativae with seven species includes the two cultivated species (*sativa* and *glaberrima*), their weedy relatives, and the wild species from which they were derived. All the Sativae carry the same basic genome, to which the symbol AA has been assigned.

Phylogenetic trends in the genus *Oryza*

Phylogenetic differentiation in the genus *Oryza* has been partially worked out by Portères (1956) and Sampath (1962). On a detailed study, rating the various characters of each species as primitive or advanced and then pooling the ratings of each section to determine their phylogenetic rank, Section Padia was found to be the most primitive and Section Oryza the most advanced (Sharma and Shastry, 1971). On close study the primitive nature of Section Padia becomes very obvious. For example, the perennial habit, absence of awn, and undifferentiated karyotype are primitive. In contrast, the elaborate sculpturing on the surface of fertile lemma and palea, well developed awn and well differentiated karyotype are indicative of the advanced phylogenetic position of Section Oryza. The karyomorphological studies of Misra (1965) and anatomical studies of Roy (1963) support the above conclusion. It is also worthy of note that every section of *Oryza* is marked by bipolar differentiation — one represented by large-seeded species and the other by small-seeded species. The latter are often associated with perennial habit, poor development of awn and greater ramification of panicle; the former carry the contrasting characters.

If the above conclusions reflect the natural phylogenetic trends in the genus, then

Table 1   *Species and subspecies of the genus* Oryza

Species or subspecies	Chromosome number	Genome	Distribution
SECTION I PADIA (ASIATIC)			
Series 1. Schlechterianae			
schlechteri Pilger	2n = ?	?	New Guinea
Series 2. Meyerianae			
granulata Nees et Arn. ex Watt	2n = 24	?	SE. Asia
meyeriana (Zoll. et Mor.) Baill.	2n = 24	?	SE. Asia
abromeitiana Prod.	2n = 24	?	Philippines
Series 3. Ridleyanae			
ridleyi Hook. f.	2n = 48	?	SE Asia
longiglumis Jansen	2n = 48	?	New Guinea
SECTION II ANGUSTIFOLIA (AFRICAN)			
Series 4. Perrierianae			
perrieri A. Camus	2n = 24	?	Madagascar
tisseranti A. Cheval.	2n = 24	?	Trop. W. Africa
Series 5. Brachyanthae			
brachyantha A. Cheval. et Roehr.	2n = 24	FF	Trop. Africa
angustifolia C.E. Hubbard	2n = ?	?	Trop. Africa (South)
SECTION III ORYZA (PANTROPIC)			
Series 6. Latifoliae			
eichingeri Peter	2n = 24	CC	E. Africa
collina (Trimen) Sharma et Shastry	2n = 24	CC	Sri Lanka
punctata Kotschy ex Steud.	2n = 24	BB	Sudan
officinalis Wall. ex Watt	2n = 24	CC	SE. Asia
australiensis Domin.	2n = 24	EE	Australia
minuta J.S. Presl. ex C.B. Presl.	2n = 48	BBCC	Philippines
malampuzhaensis Krishn. et Chandr.	2n = 48	BBCC	S. India
schweinfurthiana Prod.	2n = 48	BBCC	Trop. Africa
latifolia Desv.	2n = 48	CCDD	C. & S. America
alta Swallen	2n = 48	CCDD	C. & S. America
grandiglumis (Doell.) Prod.	2n = 48	CCDD	S. America
Series 7. Sativae			
longistaminata A. Cheval.*	2n = 24	AA	Trop. Africa
rufipogon Griff. (syn. *balunga* Sampath et Govindaswamy)	2n = 24	AA	SE. Asia
glumaepetala Steud. (syn. *cubensis* Ekman)	2n = 24	AA	Trop. America
nivara Sharma et Shastry	2n = 24	AA	SE. Asia
barthii A. Cheval. (syn. *breviligulata* A. Cheval. et Roehr)*	2n = 24	AA	Trop. Africa
sativa L.	2n = 24	AA	SE. Asia
glaberrima Steud.	2n = 24	AA	Trop. W. Africa

*   See Clayton (1968)

several further deductions can be made. It follows that the genus *Oryza* started initially as a small-sized plant growing in well drained soils in the humid atmosphere of forests. The hydrophytic habitat with preference for open sunshine and a larger size of plant (e.g. *O. officinalis* or *O. alta*) were later phylogenetic developments. The tuberculations on the surface of the fertile lemma and palea and the development of awns are also advanced characters in *Oryza,* though in the most advanced, cultivated, species *O. sativa* and *O. glaberrima* awns are often suppressed. Lastly, it follows that southeast Asia is the probable centre of origin of the genus, and its spread to the African and American continents was a later development.

Species relationships in section Oryza

To provide a background for understanding the origin of cultivated rices, it is desirable to consider the species relationships in Section Oryza. Series Latifoliae consists of diploid and tetraploid species. The diploids are distributed in the tropics of Asia and Africa, whereas the tetraploids are to be found in the tropics of America as well as of the Old World. There is nothing to suggest that the Latifoliae have contributed to the cultivated rices.

The Sativae, all of which are diploid, also have a pantropical distribution. The members of the series that are distributed in Asia and Africa fall into three types, with parallel forms in Asia and Africa. The basic type is a wild perennial grassy species which grows in ponds, ditches and canals, and which sometimes occurs as a weed in rice cultivations. It has panicles which shatter on ripening and seeds with hard seed coats and considerable dormancy. It is found throughout the Old World tropics from south China and the Philippines to west Africa and it has been recorded from Cuba and from South America.

Numerous specific names have been given to forms of this wild perennial from different geographical areas. In Asia the names *O. perennis, O. balunga,* and *O. rufipogon* have been used, and in Africa *O. longistaminata* and *O. barthii.* The New World form is usually described as *O. cubensis.* These wild perennial forms are generally regarded as representing the prototype from which the other types in the series were developed. They are closely related, though there is considerable differentiation between forms from the various geographical regions and partial sterility occurs in crosses between them.

The annual wild rices make up the second type. These forms grow in seasonal ditches. The anthers are small, the seeds are bolder and awns are prominent. The name *O. nivara* has been used for this type in Asia and *O. breviligulata* for the corresponding type in Africa. These annual wild rices are distributed over wide regions and exhibit great variation in morphology and physiology.

The cultivated rices make up the third type. These are annuals, usually growing in wet fields, but in some areas, like other cereals, on upland soils. They are upright plants with non-shattering panicles, and with seeds having permeable seed coats and a capacity for rapid germination. The cultivated rices fall into two species, namely, *O. sativa* in Asia and *O. glaberrima* in west Africa. The Asian species falls into two genetically and geographically distinct races, the *indica* race of the Indian subcontinent and the *japonica* race of Japan and north China. Between *indica* and *japonica* a range of intermediates has been identified, distributed in southeast Asia and south China regions that are geographically between the *indica* and *japonica* areas. Some of these have been treated as a separate race *javanica.* The genetical differentiation between the races is substantial, and leads to partial sterility in crosses between them. Considerable efforts have been made to extract useful agricultural races combining the merits of *indica* and *japonica,* but little success has attended inter-race hybridisation. The African cultivated rice *O. glaberrima* has also been observed as weakly differentiated into two races by Portères (1956).

The cultivated rices hybridise with the annual wild types and form hybrid swarms. These form a variable group of intermediates between the wild type and the cultivated rices. They are weedy in ecology, being found chiefly on rice fields as a crop contaminant or on field margins, annual or weakly perennial and have shattering panicles with dormant seeds. The name *O. sativa* forma *spontanea*, has been used for these types in Asia and *O. stapfii* for the corresponding types in Africa. Races that breed true have been isolated but among collections made from fields and ditches a large proportion segregate and many are partially sterile.

The nomenclature of the species included in Series Sativae is complex, as is to be expected in a crop plant, its wild relatives, and the hybrids between them. It is not proposed here to discuss the names that should be used according to the rules of botanical nomenclature, but simply to adopt a series of names by which the types under discussion can be readily identified. An unambiguous and popularly used nomenclature has been used in this paper.

The African and Asiatic elements of Series *Sativae* represent parallel situations. Each of these continents has a perennial wild species, an annual wild species and an annual cultivated species. Further, the cultivated species undergo natural hybridisation with their nearest relatives and produce hybrid swarms.

Among the wild species there has been a regular trend of evolution from perennial to annual habit, from cross-pollination to self-pollination and from lesser to greater fecundity. This is evident within the three types (Asiatic, African and American) of perennial rices as well as among the different species of the African and Asiatic continents. In this sense, it appears that the perennial wild rices are the primitive type from which the annual wild rices have evolved and the annual wild rices are the closest relatives of cultivated species.

Origins of the cultivated rices

Looking back in history and considering the evolution of cultivated rices it must be supposed that man took annual wild types (*nivara*, *breviligulata*), subjected them to the selection pressure of cultivation, harvesting and sowing and thereby gave rise to the *sativa* cultivars in Asia. The west African rice *O. glaberrima* represents the cultivated member of a series parallel to that occurring in Asia. There is no evidence of exchange between Asia and west Africa in prehistoric times to account for the existence of *O. glaberrima* and it must be accepted as a separate domestication.

The cultivated rices grow in the vicinity of wild species and offer excellent situations for introgressive hybridisation. Studies at the Central Rice Research Institute, Cuttack (Govindaswamy, Krishnamurty and Sastry, 1966) have shown that in Orissa there is introgression between these types giving rise to a wide range of forms that has been subsumed under the term *spontanea*. These forms segregate in many cases and occasionally true breeding types have also been isolated. This must have gone on from the beginning of domestication. The cultivars, therefore, arose in a situation of disruptive selection (Thoday, 1964) and the weedy types are the associates of the cultivars generated through intercrossing and occupying an ecological niche as weeds of cultivation and of field margins. Thus they have been benefitted both in ecological opportunity, and in genetic adaptation to exploit it, by their cultivated relatives.

In summary, the relationships between wild and cultivated forms may be set out as follows:

Form	In Asia	In Africa
Wild perennial	*rufipogon*	*longistaminata*
Wild annual	*nivara*	*breviligulata*
Cultivated annual	*sativa*	*glaberrima*

Introgression series

rufipogon × *sativa* ⟶ *spontanea*

nivara × *sativa* ⟶ *spontanea*

breviligulata × *glaberrima* ⟶ *stapfii*

Varietal diversity in India

It is difficult to determine precisely the time and place of origin of cultivated rice in India. Most probably, it was developed as a crop in the third millennium B.C. The earliest carbonised rice recorded from an archaeological excavation is from Lothal which has been estimated as 2300 B.C. Lothal belonged to a branch of the Harappan civilisation (Ghosh, 1961). Other finds in India have been from Rangpur (2000–1800 B.C.), Navadatoli (1550–1400 B.C.) and Hastinapur (500–300 B.C.). Rice is not mentioned in the early Vedic literature. It is highly probable that the Aryans were not aware of rice in the early stages of their colonisation of India and learnt its cultivation from the aborigines of this country. Rice does not find a place in the basic rituals of Aryans; however, it does find a place in secondary rituals which were later elaborations.

It appears, therefore, that the vast diversity of the rice crop in India has been developed in about four or five millennia. This diversity has arisen primarily in response to three major cultural factors: the range of climate in the Indian subcontinent, differences in the duration and adequacy of the water supply, and variation in the intensity of the cropping system.

It has been estimated that the total number of cultivars existing in India may be roughly around thirty to forty thousand. The total number of cultivars available at the Research Stations may be around fifteen thousand only. A small part of this material has been exploited in the breeding programme. The available material indicates that there is great diversity in the crop, not only in morphology but also in physiology and nutritional quality. The existing variability of the cultivars may be classified as follows:

1. The *indica* rices show great variation in their photoperiodic sensitivity. They have been divided into photosensitive and photoinsensitive types. The photoinsensitive types are generally early varieties maturing in 100 to 120 days. The photosensitive types generally mature under short day conditions and take more than 120 days, if they are sown at the beginning of the monsoon. Photoperiod is of great importance in determining which varieties are suitable for a given locality and season.

   The cultivars of rice are also sensitive to temperature in so far as their germination, floral initiation and maturity are concerned. However, not much has been done on varietal differences in temperature response.
2. On physiography, the rice cultivars of India can be divided into four major types:

(a) *High altitude types:* These are generally grown in the Himalayan region at elevations of 1500 m and beyond. They are generally similar to *japonica* rices and are resistant to cold.

(b) *Upland types:* Rice is grown generally as an upland crop in those large areas of India with a rainfall of 750 mm or less. Upland rices are generally short-duration varieties and resistant to drought.

(c) *Lowland types:* The major part of the rice-growing tracts in India is occupied by the lowland types which contribute most towards total production. These are generally long-duration (photosensitive) and are grown on wet land.

(d) *Deep-water types:* The area occupied, and the number of varieties, in this group are very limited. Areas which are submerged during the rainy season and accumulate water to a depth of 1−10 m grow a type which elongates with the rising level of water. The major part of the plant floats on or near the surface of the water.

3. Morphological variation also plays a great role in varietal classification. The grain type is probably the most important factor in this connection. Commercial rice varieties are classified as fine, coarse or medium depending upon the size of the grain. The fine types in India sell at a premium and are preferred for consumption. Besides, the size of the grain, leaf colour (green or purple) the kernel colour (red or white) and kernel type (translucent or opaque) also play an important role in commercial classification.

History of rice breeding in India

Early work on rice improvement was confined to pure line selections from local varieties. Some of the improved varieties thus identified by eminent workers like Hector, Graham, Parnell and Ramiah are still popular at the present time. In a few cases, hybridisation was resorted to, for incorporation of disease resistance. While this was the best that could be done during the period of 1900−50, the procedures of breeding did not permit nationwide testing and this led to the proliferation of rice varieties 'suited' to restricted areas. Further, sharing the defects in plant type with other countries of tropical Asia, these improved varieties were prone to lodging and were poor in nitrogen response.

The credit for identifying the major deficiency in tropical rices should go to Dr K. Ramiah, who formulated a scheme of hybridisation between fertiliser responsive *japonica* varieties and tropically adapted *indica* varieties. This project was indeed the genesis of a national programme and even fostered regional co-operation between the countries of southeast Asia. Although much was expected of this programme, the results were disappointing, as the theoretical implications of such inter-racial hybridisation were not clearly understood, and the breeding techniques required for success were not worked out. The only variety originating from the programme which was grown on a considerable area was ADT 27.

The yield potential and adaptability of dwarf rice varieties of the *indica* type was highlighted by the International Rice Research Institute (I.R.R.I.). The ideas on plant type that originated from Japan were realised with *indica* rice by the use of two varieties, Taichung (Native) 1 and Dee-geo-woo-gen from Taiwan. The Institute undertook a massive hybridisation programme between these dwarf varieties and tall varieties from different countries. Indian scientists introduced the dwarf variety Taichung (Native) 1 from I.R.R.I. during 1964 and undertook a programme of hybridisation between local varieties and dwarf rices. The programme at I.R.R.I. identified IR 8 which has been released in India after testing. Subsequently, the Indian programme evolved several dwarf varieties, the most important of which are Jaya and Padma.

The new era in rice breeding associated with the dwarf rice varieties involves more than the breeding of short-strawed types. Success in plant breeding depends on breeding to exploit the opportunities offered by the agronomic system. In this instance the new opportunity which the dwarf rices exploit is the opportunity to establish a much higher fertility level in the rice lands by the use of fertilisers. With higher fertility levels there is a need for better water management and good control of pests and diseases. And the ability of the dwarf varieties to respond to high fertility with heavy yields, and without lodging, makes it profitable to make use of the extra inputs.

The two introduced dwarfs Taichung (Native) 1 and IR 8, showed the potential of the new era. This potential will only be realised by a sustained and co-ordinated breeding programme in India that will result in nothing less than a new evolutionary advance in the rice crop. Varieties must be bred for the whole range of cropping circumstances. Dwarf rices are of no use for deep water conditions, for example, and a range of varieties is needed for cropping systems of different duration. Quality demands are important, and varieties are already available with the characteristics of some of the traditional high quality Indian varieties. The range of types needed is somewhat reduced by the availability in the dwarfs of photoinsensitivity. Finally, the importance of pest and disease resistance is very great indeed, and is under study at all breeding stations.

The strategy of the rice breeding programme has been developed greatly in recent years. With a closely co-ordinated system for the exchange of information and of material, a breeding programme covering the subcontinent is planned to exploit the genetic resources of the *indica* rices, and to put through trials and into production the best of the lines that emerge.

Job's tears

A.K. KOUL

P.G. Department of Botany, Kashmir University, Srinagar

The generic name *Coix* is of Greek origin (Vallaeys, 1948) and it occurs in the works
of Theophrastus, who in the fourth century B.C. applied it to a 'reed-like' plant
which might have been a form of modern *Coix*. It appears that the plant described
by Pliny in the first century A.D. and named *Lithospermum* was also a form of
Coix. The common English name is 'Job's tears'. The expression as applied to *Coix*
rests on the shape of the grains. Gerarde wrote 'every graine resembleth the drops
of teare that falleth from the eye'. The Arabs called *Coix* seed *'damu-Daud'* or
'tears of David' and later *'damu-Ayub'* or 'Job's tears'.

 Coix is one of the eight genera included in the tribe Maydeae (Family Gramineae)
of Engler and Prantl (1889). Of this tribe, three genera (*Zea, Euchlaena* and
Tripsacum) belong to the New World and five (*Sclerachne, Trilobachne, Chionachne,
Polytoca,* and *Coix* are of Old World origin. Anderson (1945) and others have made
attempts to link *Coix* with the ancestry of maize, but the fact is that the differences
between the Oriental and the Occidental genera outweigh the similarities. Apart
from differences in the place of origin, the two groups exhibit great differences in
morphology and cytology. All these led us (Koul and Paliwal, 1965) to split the
heterogeneous tribe Maydeae into two homogeneous tribes: (1) Maydeae, including
the American genera, and (2) Coixideae, including the five Oriental genera.

 Apart from morphological descriptions, little is known about *Chionachne,
Sclerachne, Trilobachne* and *Polytoca*. Morphologically *Coix* differs from all four
in the nature of the involucre enclosing the female flower. In *Coix* the shell is formed
by the hardened leaf sheath while in the remaining four genera the covering represents
a hardened glume or glumes.

 Coix is widely distributed in southern and eastern Asia and some forms are now to
be found almost throughout the world. Most species of the genus are, however, met
with in India, Sikkim, Burma, the Philippines, Japan, Malaya, Indonesia and Sri
Lanka. Arab travellers in the east became acquainted with the seeds of *Coix*, and they
are held responsible for its introduction into Spain and Portugal, where it is
naturalised and is still known as *Lagrima de Job*. From here the plant spread to
other parts of the west, primarily as a garden curiosity.

 Though a small and well defined genus, *Coix* is an assemblage of ill defined and
poorly understood taxa, largely because of the numerous intergrading forms. Thus
agreement regarding their status has yet to be reached. Of more than twenty taxa
that have been described, only five or six deserve specific rank. The established
species of the genus are: *Coix aquatica, C. poilaeni, C. lacryma-jobi, C. puellarum,*
and *C. gigantea*. Species of *Coix* form a polyploid series with base number 5

(Darlington and Wylie, 1956). *C. aquatica* and *C. poilaeni* are diploid (2n = 10) (Mangelsdorf and Reeves, 1939; Nirodi, 1955). All the varieties of *C. lacryma-jobi* are tetraploid (2n = 4x = 20). *C. gigantea* occurs in tetraploid and octoploid forms (2n = 8x = 40). All these species constitute a euploid series. The only known taxon of the genus with an aneuploid number is that described by Koul and Paliwal (1964) with 2n = 32.

On morphological grounds, Watt (1904) recognised two series within the genus. The 'aquatica–gigantea' series includes species with linear lanceolate leaves (30–60 cm long and 5–7 cm wide) bearing hairs arising from pale, crateriform glands. To this series belong *C. aquatica, C. poilaeni, C. gigantea* and the aneuploid species of Koul and Paliwal (1964). The 'lacryma-jobi' series includes forms with wider (11–13 cm wide) and comparatively less hairy leaves. These series represent the two lines along which evolution has occurred within the genus.

Members of the aquatica–gigantea series are 'always wild plants never brought under cultivation and found growing on the lower hills, dry soils and swamps' (Watt, 1904). In this series the capsular spathe is stony hard, pyriform, much drawn out at the apex with the mouth cut obliquely into an elongate serrulate lip; ripe fruit highly polished, prominently angled having two or three furrows along its flattened face, of a dull greyish, white or brown colour. These species are useless as food plants on account of the extreme hardness of the fruit. They have a restricted distribution. *C. gigantea* is a pernicious weed of paddy fields.

The lacryma-jobi series includes the wild and cultivated forms of *C. lacryma-jobi.* This series is represented in India, Burma, China, Japan, Malaya, Sri Lanka, West Indies, Polynesia, the Mascarene Islands, the American continent, tropical and north Africa and southern Europe. There are three or four well marked forms of Job's tears met with in India which differ from each other in shape, colour and degree of hardness of the involucre. These forms have been raised to the status of varieties by Watt (1904). The four varieties are: var. *typica*, var. *stenocarpa*, var. *monilifer* and var. *Ma-Yuen.*

On the distribution of *Coix* in India and adjoining regions Watt (1904) wrote '...*Coix* would have to be commented on as a feature of great interest in the tract of country that stretches east by south from Nagpur to Sikkim, Assam, Burma, Malaya and China...'. Vallaeys (1948) suggested the Malay Archipelago as the centre of origin of *Coix,* taking into consideration the multiplicity of varieties present there, some of which are not found elsewhere in the wild state. It seems to have early become a cereal of some importance in hilly regions of southern China (Burkill, 1935) and therefore the beginning of cultivated races should be sought in that part of the world. The existence of an extensive series of cultivated forms and the occurrence of a long list of names for the plant and the grain in nearly every vernacular language of India, Burma, Malaya and China are indications of its early cultivation in this region of the world. According to Pieris (1936) the edible grain (known as *adlay* in the Philippines) has been in cultivation in India for some 3000–4000 years. Cultivation is fairly extensive and widely dispersed among aboriginal tribes. There are great diversities in size, shape and colour of the grain, as also in the quality and the purpose to which it is put. These diversities, and the existence of many vernacular names, confirm the belief that it has been known from ancient

64

times. Of Watt's four varieties of *Coix lacryma-jobi,* var. *Ma-Yuen* alone is of importance as a source of human food. The varietal name (Ma-Yuen) is after the Chinese General who conquered Tongking in the first century A.D. and who was so impressed by the Bo-bo (*Coix*) grain of that place that he carried cartloads of seed and introduced and popularised it in China. All other forms of the species are useless as food plants, primarily on account of their polished, stony hard fruit spathe which makes hulling very difficult.

The hard-shelled forms have long been used in the orient for ornamental purposes in rosaries, necklaces, curtains and draperies. In fact, *Coix* caught the attention of Westerners not so much as a food plant but as a garden curiosity. Attempts were made to improve the hard-shelled forms of *Coix* to make them more attractive and abundant. An investigation of the forms of *Coix* in India and Burma was instituted towards the end of last century at the instance of W.T. Thiselton Dyer, the then Director of the Royal Botanic Gardens, Kew, to determine the quality and degree of variability particularly with regard to the shape, colour and size of the fruit shell. Cultivation trials with the hard-shelled varieties of *Coix* were abandoned because it appeared that 'cultivation destroys very rapidly the hard pearly shell, upon which to a very large extent the industrial demand depends. It also changes the colour of the grain and produces chalky white and straw colours utterly devoid of the rich glossiness of the wild grains.'

The grain of *Coix* is large, and this is a factor which should have favoured its use over other cereals as a food plant but for the stony, shining involucre which made hulling difficult. Thus the cultivated races had to be selected for easy husking. In the forms specially cultivated the shell is soft and amenable to ordinary methods of milling. Though the edible grain is almost unknown to the inhabitants of India generally, to many of the aboriginal races such as those of Nagaland, Assam, Sikkim, etc., it is an important article of diet. Watt (1889) believed that *Coix* may have preceded rice and in the plains of India may have been abandoned in favour of the more wholesome grain. *Coix* is also cultivated for food in Sri Lanka, the Philippines and Brazil.

Apart from its large seed, sweet taste and easy husking the variety Ma-Yuen offers several other characters that increase its scope as a cereal. These are: (1) it can be used in the preparation of any article of food usually made of rice and also with the same degree of palatability. (2) It can be grown successfully in areas not suited for other crops. Cultivation processes are also quite simple. (3) It is less subject to insect and fungal attacks, and above all, (4) it is more wholesome than wheat and rice, as it contains a greater proportion of fat and protein (Table 1).

The interest of *Coix* in studies of crop plant evolution is in the limited extent of its development. It appears to have been domesticated early, and the major disability of its hard shell was overcome in the Ma-Yuen variety. Nevertheless it remained a crop of poor cultivation by primitive peoples, used as a reserve against periods of scarcity and not as a major food. Thus such primitive characters as long growing season, uneven ripening and variable yield, have never been overcome by plant breeding, either by farming peoples or by research scientists; yet there seems no reason to doubt that varieties suited to modern agriculture could be bred. There is great diversity in the species, and indeed a dwarf, high-yielding variety has been

Table 1   *Chemical analysis of* adlay, *wheat and rice* (after Pieris, 1936)

	Water	Protein	Fat	Carbohydrate	Fibre	Ash
Adlay (threshed)	10.8	17.6	5.6	62.1	0.3	3.6
Rice (polished)	12.2	7.6	1.0	77.9	0.3	1.0
Wheat	10.6	12.2	1.7	71.1	2.3	1.8

bred in Brazil (Schaaffhausen, 1952). It seems likely that the development of *Coix* ceased when the greater merits of rice were recognised, and the prospects of continuing the improvement of *Coix* will depend on whether there arises in world agriculture a situation in which *Coix* has an advantage over other cereals. The potential is there if at any time it is needed.

Rape and mustard

A. NARAIN
Division of Genetics, I.A.R.I., New Delhi

Brassica, the genus to which rape and mustard belong, is most versatile and has contributed a large number of species of great economic value. It comprises about 159 species of diverse habit, mostly native in the north temperate and Mediterranean regions of the Old World. Almost all parts of the plant — root, stem, leaves, inflorescence and seeds — contribute important products in one or another species of the genus. Both diploids and polyploids have contributed species of great agricultural worth.

Harberd (1972) has recently given a detailed account of the cytotaxonomy of *Brassica* and some closely related genera. He has grouped the species of the genus into 'cytodemes', within which the basic chromosome number is constant and the species are cross fertile, and between which the basic chromosome number often differs and cross sterility is the rule. Of the species of interest in agriculture, four are essentially diploid, and three are allopolyploid. Each belongs to a separate cytodeme. The diploids are: *B. nigra* (n = 8, B genome), *B. oleracea* (n = 9, C genome) *B. campestris* (n = 10, A genome) and *B. tournefortii* (n = 10, D genome). There is evidence (Robbelen, 1960; Narain, 1969) that the 8, 9, 10 series represents secondarily balanced diploids derived from a basic $x = 6$. The three allopolyploid species have been derived from these diploids; *B. napus* (n = 19, AC genomes), *B. juncea* (n = 18, AB genomes) and *B. carinata* (n = 17, BC genomes). Two amphidiploids have been synthesised experimentally: synthetic *juncea* (AB from *B. campestris* × *B. nigra*) and a near-*juncea* (DB from *B. campestris* × *B. tournefortii:* Narain, 1967).

The Indian cultivated Brassicas are oleiferous types belonging to two species, *B. campestris* and *B. juncea.* The use of *Brassica* species for vegetables, which are of enormous dietary significance both in Europe and in eastern Asia (see Herklots, 1972), is unimportant in India. *B. campestris* has given rise in India to three distinct cultivars: brown *sarson,* yellow *sarson* and *toria.* They are cultivated in rather distinct ecogeographical areas, and may be regarded as ecotypes. Collectively they are known in India as rape, and they constitute an important oilseed crop. *B. juncea,* or *rai,* is a more vigorous cultivar, and is also an important oilseed crop. It is known in India as mustard.

Origin and history

The cultivation of rape and mustard goes back to Harappan times (Pigott, quoted by Mehra, 1968). They were in use in the Vedic civilisation. References to use, whether in medicine or religious ceremonies, are found in ancient Sanskrit literature

(Mehra, 1968). The fact that there is hardly any reference to these oleiferous Brassicas in any other contemporary civilisations, suggests their early domestication in the country. 'Rajak' has been used for black mustard, *B. juncea* (Watt, 1889), and 'Siddhartha' for the white rape which is *B. campestris* (and not *B. alba*, see Prain, 1898), particularly yellow-seeded *sarson* which because of its attractive colour was classified separately.

In discussing the origin of the oleiferous rapes and mustards, the rapes, *B. campestris,* are diploid and must be considered first. *B. campestris* occurs wild as a weed from western Europe to eastern China. Thus the Indian rapes fall geographically midway in the vast geographical distribution of the species. The wild, unspecialised primitive rape became a weed of cultivation, and gave rise to a range of cultivars as a result of three different types of selection in the three main regions of its habitat. In the west, selection for root development gave rise to the turnip. In the Far East selection for leafy vegetables yielded the great diversity of Chinese cabbages that are classified as *B. pekinensis* and *B. chinensis* (Herklots, 1972). In the mid-region, in India, selection for oil content gave the three oleiferous races, brown and yellow *sarson* and *toria*. All these types are cross compatible and they belong, in Harberd's terms, to the same cytodeme. The Indian oleiferous group appears to have developed by the differentiation of the brown *sarson* stock. According to Russian workers, and also Singh (1958), eastern Afghanistan and adjoining areas of Pakistan and northwest India may be regarded as its centre of origin. From this, yellow *sarson* arose by selection of yellow-seeded types that were regarded as of superior quality. *Toria* arose in response to selection for adaptability to the ecological situations of Bengal, Bihar, Orissa, Uttar Pradesh and Punjab.

Turning now to mustard, *B. juncea* is an allopolyploid carrying the genomes of *B. campestris* and *B. nigra*. The diversity of the species has been discussed by Vaughan, Hemingway and Schofield (1965), and by Herklots (1972). Vaughan *et al.* record that 'It is grown particularly in Eastern Europe, India, Pakistan, China and Japan and has been introduced into other countries ...'. Herklots describes the wide range of 'mustard cabbages' of this species that are grown in southeast Asia. Most are grown there as green vegetables for leaves and stems, but there is one, var. *megarrhiza*, that is turnip-rooted. The Indian and eastern European forms are grown as oilseeds.

B. campestris and *B. nigra* are sympatric over the greater part of the range of *B. juncea,* and the allopolyploid may have arisen almost anywhere from eastern Europe to China. Vaughan *et al.* (1965) have shown that *B. juncea* cultivars can be classified in terms of glucoside content. With only very few exceptions the seeds of northern forms — from eastern Europe, Russia, China and Japan — contain glucosides that give rise to allyl isothiocyanate. Seeds of Indian and Pakistan forms contain glucosides that produce both allyl and 3-butenyl isothiocyanate and occasionally 3-butenyl isothiocyanate with no allyl isothiocyanate. Vaughan *et al.* added morphological characters to these biochemical characters and divided *B. juncea* into two geographical groups:

(1)   Indian and Pakistan forms — little variation in leaf form, only brown seeds epidermis normally absent in the testa, 3-butenyl isothiocyanate produced.

(2)   Other Asian forms (China probably the main centre of diversity) – wide variation in leaf form, brown or yellow seeds, variation regarding presence or absence of the epidermis, only allyl isothiocyanate produced.

Herklots' (1972) account of the vegetable forms of eastern Asia confirms the view of Vaughan *et al.* that China is a major centre of diversity, and shows that the wide variation in leaf form is a product of the selection for use as a green vegetable. Vaughan *et al.* go on to suggest that the diversity in *B. juncea* may have arisen in part from multiple origin of the allopolyploid, with several different members of the *B. campestris* cytodeme involved. This possibility has received support from Narain's (1971*b*) work on the synthesis of artificial amphidiploids. He has made it clear that whereas oleiferous *B. campestris* has contributed as one of the constituent parents of oleiferous *B. juncea*, leafy *B. campestris* sub sp. *narinosa*, etc. has contributed in the same way to leafy *B. juncea* (e.g. var. *cernua*).

Brassica tournefortii, a diploid with n = 10 chromosomes, is a wild or spontaneous species in northern India and westward through west Asia and the Mediterranean to Italy and Spain, where it occurs as a weed (Olsson, 1954). It is said once to have been cultivated in Rajasthan and Punjab (Prain, 1898; Hooker, 1872). Oil must have been the chief attraction for its cultivation. The present type contains nearly 30 per cent oil. At present, it is only found growing spontaneously in these states. There is hardly any mention of this crop in ancient Indian scripts. It is reported to be under cultivation in Tibet. It is well adapted to a dry habitat and ill adapted to high humidity, which may be the reason for its not having spread eastward of Punjab in the Indian peninsula. It also explains its disappearance from cultivation in Punjab with the introduction of canals in the area. It is self-fertile and sets seeds freely when isolated in paper bags. Until recently very little was known about its genomic relationship with other species of *Brassica*. However, the work of Narain (1968) based on chromosome pairing in interspecific hybrids has thrown much light upon its ancestry and origin. *B. tournefortii* is karyotypically very similar (Sikka, 1940) and morphologically very near to the species of the *B. campestris* cytodeme, but it is not freely cross compatible with it, and hybrids when made are sterile, and allow very little gene flow between the species. Nevertheless, sufficient gene exchange can take place for this species to be regarded as a source of aphid resistance in breeding for improved oleiferous rape (Singh and Narain, 1965). Narain (1968) regarded the genomes as distinct from the *B. campestris* (A) genome, and has assigned it the letter D. Harberd (1972) regards it as a separate cytodeme, in conformity with Narain's views. Hybrids generally produce twenty univalents, but bivalents are produced in small numbers. It is interesting to note that in crosses with the three main types of *B. campestris*, fewer bivalents are noted in the hybrid involving the leafy (0.21 mean) and rapiferous groups (0.40 mean) than the oleiferous group (0.90 mean). On the pairing relationships in meiotic metaphase, it may be inferred that *B. tournefortii* is more closely related to the oleiferous group than to the other groups of *B. campestris*.

It may be supposed that a common ancestor of the oleiferous type differentiated into two stocks somewhere along the line from the Mediterranean to northwest India. The *B. campestris* line proved more successful, and more adaptable, and gave rise to the three oilseed types, brown and yellow *sarson* and *toria* that have met the ecological situations of the north Indian oilseed areas. The *B. tournefortii* stock, on the other hand, remained adapted to the limited drier habitats of northwestern India.

Genetic improvement of rapes and mustard

Genetic improvement of rapes and mustard has been going on since the second decade of this century in the major growing states. For the improvement of yield, pure line selection in the self-pollinated group (*B. juncea,* yellow *sarson* and the tora type of brown *sarson*) and mass selection, along with the production of a short-term synthetic, for local adaptation in the cross-pollinated group (*toria,* and lotni type of brown *sarson*) were followed up to the early 'forties. In later years, during the 'fifties, these breeding methods were supplemented with induced polyploidy, and mass pedigree selection and mutation breeding, particularly in the improvement of *toria.* The improved types so evolved, although registering 10–20 per cent higher yield than the land varieties, did not spread satisfactorily. In the first three Five Year Plans starting from 1950–1, increased production was realised mostly from improved agronomic practices and increasing the area under cultivation. In the subsequent plan, the research projects on these crops were intensified and launched on an All-India Co-ordinated level.

The 'green revolution' has hardly touched the *Brassica* oilseeds. The breeder has not yet succeeded in advancing far beyond the yield potential of the land races. Most breeding work has been conducted within the limited range of the Indian races of the two oilseed species. There has until recently been little critical analysis of plant type in relation to yield synthesis, and such work as was done revealed close linkages between physiological characters, and consequently a severe limitation on the breeder's ability to synthesise new, high yield potential character combinations.

To overcome these limitations, new breeding techniques were applied in the 'sixties in breeding programmes with brown *sarson* and mustard. In brown *sarson* Murty, Arunachalam, Doloi and Ram (1972) were able by disruptive selection to break down character associations and to release variability and recombination potential in the components of yield. In mustard it has been claimed (Narain, 1971*b*) that much potentially useful variability has been released in crosses with amphidiploids artificially synthesised from the putative parents of that species. Though showing considerable promise, these programmes have not yet produced new varieties of outstanding performance in the All-India Co-ordinated Trials.

Further physiological studies have also been undertaken. The inter-relationships and heritability of yield and its components have been analysed in brown and yellow *sarson* and in mustard. Yield depends primarily on the length of the main stem, the number of branches, the number of seeds per pod and seed weight. Total number of pods per plant, number of seeds per pod and seed weight were shown to be highly heritable and to give a high expected genetic gain in selection for high yield. These studies were based on the spectrum of variability available in Indian research stations only, and the conclusions might be modified if a wider range were examined. Nevertheless, they suggest that breeding progress may result from re-structuring the plant type by selecting for a greater number of branches, larger and more densely podded main stem, and more and heavier seeds per pod.

Castor

A. NARAIN
Division of Genetics, I.A.R.I., New Delhi

History and origin

The castor plant, belonging to the monotypic genus *Ricinus,* has been under cultivation since the dawn of human civilisation. Its generic name, *Ricinus,* derived from
a Latin term meaning 'Dog tick', was given to the plant by Linneaus because of the
resemblance of its seeds to the canine pest. The word castor was coined by English
planters from Agro Casto, the name which was in common usage by Portuguese and
Spaniards in Jamaica where it was extensively cultivated during the eighteenth
century. Castor is a native of the tropics and is grown under very diverse conditions
of soil and climate. It is generally adapted to a well drained sandy loam soil receiving
38–50 cm (15–20 inches) rainfall. It is usually a crop of rain-fed areas but the
increased yields from improved varieties and hybrids have extended the area of
economic adaptation even to irrigated areas.

The original home of the plant is uncertain. Hooker (1890), de Candolle (1886)
and Hilterbrandt (1935) held that the evidence was in favour of an African origin,
whereas Fluckiger and Hanbury (1892) were of the opinion that it originated in
India. De Candolle (1886) found it wild in African countries. Watt (1892) reported
that it is found growing in the wild state in the scrubby jungles of the outer Himalayas.
Describing its antiquity in India, he observed 'It thus follows that the use of the
castor oil plant was known and the plant was very probably cultivated in India many
centuries before the Christian era.'

Castor seems to have been introduced in agriculture sometime when man was
learning to tame light. The earliest record of the utility of its seed is found in
Egyptian monuments and in ancient Sanskrit literature. In prehistoric times, its oil
was mainly used for illumination and for medicinal purposes. With the development
of human civilisation, particularly in the period that followed the Industrial
Revolution, it found diverse uses in industry.

The cytology of castor

Richaria (1937) studied secondary bivalent associations and reported that castor
is a secondary balanced polyploid with the genetic constitution **AAA BB CC DD E**
($2n = 20$). Subsequently similar associations were noted by Kurita (1946) between
two pairs of bivalents and by Jacob (1957) between pachytene chromosome E and
G. The ploidy status of the castor plant was finally established by the isolation by
the author of a spontaneously occurring haploid castor plant and the study of its
meiotic behaviour at diakinesis and metaphase I. Five bivalents were observed both
at diakinesis (Plate 1*a*) and metaphase I (Plate 1*b*) and five groups each with S–S

PLATE 1
Meiotic stages in a haploid castor plant. (*a*) Diakinesis showing 5II.
(*b*) Metaphase I showing 5 II. (*c*) Metaphase I showing five groups with
S–S pairing.

chromosome pairing (Plate 1*c*). Thus there is evidence that the contemporary species with $2n = 20$ arose by autoploidy from a diploid progenitor species with $2n = 10$. The basic number of the species is, therefore, $x = 5$ (Narain, 1968) and not $x = 10$ as reported by Darlington and Wylie (1956). Moreover Richaria's constitution should be amended to $n = $ AA BB CC DD EE.

Genetic diversity in castor

The castor plant is very polymorphic in nature. The diversity of its forms attracted many botanists in the past to attempt a classification of the range of its variability. Most of these classifications appeared to be either artificial or ecogeographical in principle. The need of a breeder for a scheme describing the variability on some natural order can hardly be over-emphasised. Such a system should obviate any duplications which are likely to occur due to similarities and transitional features in intra-ecogeographical groups and should embody a clear and ready account of the variability in the species. An effort is here made to describe the range of diversity in some important characters of the plant in what may represent an evolutionary sequence.

Habit of the plant

The castor plant exists in both perennial and annual forms. Perennial types form large shrubs or small trees. Annuals are generally herbaceous. The size of the plant is largely determined by its height, which depends in large measure on internode length and number of nodes to the first inflorescence. Narain (1958*a*), working with the type collection at Kanpur, found the node number ranged from 24 to 10 between types; in any one type the number was fairly constant and was characteristic of that type. In tropical conditions, perennial types with a high node number may grow to a height of 9–12 m., with stems 7.5–15 cm in diameter. Annuals usually grow to 1–2 m in height. The perennial type is primitive. Annuals are of recent origin, having been selected for their suitability to modern agriculture.

Colour of vegetative parts

The primitive form of the castor plant was probably green throughout. Anthocyanin pigmentation has arisen, and is particularly well developed in an extensive series of coloured ornamental forms. Pigmentation varies both in intensity and in distribution. Following Ridgway's (1912) colour charts, the following intensity range can be distinguished: dark green, light green, rose pink, sun red, deep rose pink, rose colour, rose red, pomegranate, Bordeaux, and burnt lake. Colour patterns are best set out in tabular form, as in Table 1.

Among the agricultural types, both green and coloured are common. The different shades of colour are neutral to selection pressure in the development of characters of agricultural excellence. However, they have played an important role in the evolution of the finest ornamental types (Narain, 1960).

Bloom character

Many castor plants develop a waxy bloom on the plant body. The intensity of the bloom varies, and it may be absent altogether. It is easily rubbed off. Various types

Table 1　*Colour patterns in ornamental castor plants*

Pattern	Stem	Petiole	Leaf		Capsule			Spines
			Lamina	Veins	Main	Ridges	Stigmatic end	
Green	−	−	−	−	−	−	−	−
Red spines	−	−	−	−	−	−	−	+
Red spines +	−	−	−	−	−	−	+	+
Red spines ++	−	−	−	−	−	+	+	+
Red stem	+	+	−	+	−	+	+	+
Red capsule	+	+	−	+	$\frac{1}{4}-\frac{3}{4}+$	+	+	+
Red	+	+	+	+	+	+	+	+

Anthocyanin absent : −　　　　　　　　Anthocyanin present : +

Table 2　*Distribution of waxy bloom on castor plants*

Class	Bloom classification	Areas covered					Intensity
		Stem	Petiole	Leaf		Raceme	
				Under surface	Upper surface		
1	Treble	+	+	C	C	+	high
2	Partial treble	+	+	C	P	+	high
3	Double	+	+	C	−	+	high
4	Partial double	+	+	P	−	+	high
5	Single	+	+	−	−	+	high
6	Smoky light	+	+	−	−	+	low
7	Raceme	−	−	−	−	+	low
8	Nil	−	−	−	−	−	−

C = complete cover
P = partial cover

of bloom have been differentiated by Narain (1952). It provides a good adaptation to the plant for survival under moisture stress, and the presence of bloom appears to be a primitive character. Long domestication of the plant, and cropping under conditions of assured moisture, have resulted in the evolution of various bloomless types. The progressive disappearance of bloom from the vegetative parts of the plant, indicating the stage of evolution under domestication on the basis of bloom, is shown in Table 2. Bloom and anthocyanin pigmentation are independently assorted in the field and the horticultural populations.

Flowering pattern, earliness, and crop duration
The main stem of the castor plant terminates in an inflorescence. Thereafter, two vegetative nodes are produced on each of the uppermost laterals, followed by a bud which has the potential to grow into an inflorescence. This pattern is repeated throughout the growth of the plant. There is therefore a geometric pattern of

location of potential flowering points. Actual flowering is very much less than the potential since only a few of the buds at the potential flowering points actually sprout, and of these only a few develop into inflorescences. The number and distribution of laterals that develop, and of potential flowering points that produce inflorescences, determine the structure of the plant. Sethi (1931) and Masur (1933) have described two main types of branching: sparse and bushy, or top and basal. Plant habit does not in fact fall into two sharply defined classes. An intermediate pattern of branching is the common type. Strict basal branching is to be found. Full top branching only occurs on a few types, on dwarf annuals more commonly than on perennials.

Recent breeding work has been directed to the development of early annual types with a plant habit suited to close spacing to enhance yield per hectare and synchronous fruiting and indehiscent capsules to facilitate harvesting. The node at which the first inflorescence appears is an index of earliness. This is readily recorded in the field, and is responsive to selection. The maturity period, from first inflorescence to completion of ripening, is compounded of the duration of flowering and the time from flowering to seed ripening. Unimproved, long term annual cultivars have an extended flowering period. Seed is harvested by repeated pickings, and the total cropping period from seeding to completion of harvest may be 250 days. Short-term annuals have been selected which produce the first inflorescence at a low node, and have a top branching plant habit. Among these are types with condensed internodes and synchronous sprouting of the inflorescence-producing buds. The capsules mature over a short period, and when combined with the non-dehiscent capsule character, the crop can be harvested at a single pick, with a total crop duration of 120–150 days.

Types of inflorescence

The normal inflorescence of the plant as described by Chandrasekharan *et al.* (1946) is a monoecious panicle having pistillate flowers on the upper and staminate ones on the lower axis of the raceme. The number of fruits borne on the main axis of the inflorescence, and the compactness of the inflorescence (fruits per unit length, Masur, 1933) vary greatly. Selection for agricultural productivity has favoured high density in the form of compact pyramidal or cylindrical head types. The partition of the panicle between male and female flowers also varies. Classen and Hoffman (1950) recorded the following variations from the standard pattern, giving racemes with:

1. pistillate and staminate flowers interspersed
2. 70 to 99 per cent pistillate flowers
3. 100 per cent pistillate flowers
4. hermaphrodite flowers

Among the cultures studied by the author, besides all the types of inflorescence described above, one had a central raceme with staminate flowers and secondary racemes with 1–2 per cent pistillate flowers.

The identification of 100 per cent pistillate plants and plants with 100 per cent staminate racemes shows the prospect of progression from the monoecious to the dioecious plant. Breeders are at present engaged in improving the stability of pistillate lines and in increasing the number of racemes per plant that they produce. These characteristics are of recent origin and are of value for the exploitation of the hybrid vigour so often manifested in the crop (Zimmerman and Parkey, 1954; Narain, 1958a, b). The evolution of some recent varieties has been made possible by the existence and exploitation of pistillate forms.

Types of fruit

The fruits of castor plants may be spiny, spineless or sparsely spiny. Narain (1951) reported a mutant which in its raceme bore different types of capsule characterised by some with spines, some with partial spines, some with very sparse spines and some with not a single spine throughout. The secondary inflorescences of the mutant were either completely spiny or spineless. The length of spines also varies considerably in different types, ranging from very long spiny types to short ones. The character spininess is of evolutionary significance. It once had an adaptive value to a wild plant but became an undesirable character when it was brought under protective cultivation. Spineless types, though much sought after in agriculture, always have loose racemes and hence are not generally associated with high yield.

The fruits of the wild plant are dehiscent, splitting violently and scattering the seeds when ripe. Dehiscence was an important adaptive character associated with survival in the wild. In cultivation it is disadvantageous, leading to substantial losses at harvest. Indehiscent types occur, and are being exploited in the breeding of the productive annuals required in modern agriculture. Indehiscent types generally predominate in succulent spineless forms and forms with thick-walled capsules. They have been subdivided (Narain, 1959) into:

1. types with incomplete septicidal dehiscence
2. types with incomplete septicidal and loculicidal dehiscence.

Seed

The size of castor seed is very important as it is the standard by which castor is judged in the trade. There is a great variation in the dimensions and weight of the seed. Some small-seeded varieties have seed less than 1 cm long, running about 10 000–11 000 to a kg while the large Zanzibar ornamental types with seed more than 2.5 cm long run only 990 to a kg. The seed also differs in shape as some are oblong, some oval and still others orbicular.

The size of the aril or caruncle on the ripe seed also shows much variation. Hilter-brandt (1935) classified them as:

1. Persian type with no caruncle
2. Chinese type with small caruncle
3. Indo-African types with well developed caruncles

As the caruncle is the development of the third integument, its absence in the type led to a thorough investigation of the character starting from early unmatured to well developed matured seed. It was found that in aril-less types, the caruncle is quite well developed in early stages but in the course of ripening of the seed, the caruncle dries up and the seed takes on the appearance as if there is no caruncle. Such types, may, therefore, be described as pseudoaril-less types.

The oil content in the seed ranges from 44 to 56 per cent depending upon the nature of the variety. Experiments conducted at Hyderabad show that the type of soil has no bearing on the oil content of the seed. It is, however, affected to some extent by the degree of maturity and the completeness of filling of the seed. Oil percentage is not related to spininess of the fruit, anthocyanin pigmentation, or bloom (Narain, 1962).

In order to see whether there is any relationship between seed size and oil content, forty varieties were analysed. In the small-seeded group, the average oil content was 50.0 per cent ranging from 43.4 to 55.7 per cent, and in the bold-seeded group 51.2 per cent ranging from 44 to 56.6 per cent, and thus there does not seem to be any association between the oil content and seed size amongst the varieties studied.

Conclusions

Summarising the information set out above on the characters of the castor plant, the evolutionary development of the species may be set out as follows: Primitive castor was completely green, perennial, with a bloom, and arboreal in habit. Its inflorescence was loose, fruits very long-spiny, dry, dehiscent, thin-walled, the seeds with a well developed aril. In the process of domestication two distinct lines emerged, first the agricultural type grown, and improved by selection, for the oil content of its seeds, and later the horticultural stock, selected for ornamental characters. The modern agricultural type is dwarf, annual and herbaceous in habit. Its internodes are very condensed so that flowering branches sprout more or less synchronously. The fruits are indehiscent, sparsely spined, compact, maturing very early and almost all at one time. The characteristic feature of the horticultural race is its diversity in habit and duration, and in the distinctively ornamental characters; anthocyanin pigmentation, bloom, and inflorescence and fruit type.

Pigeon pea

D.N. DE

Applied Botany Section, Indian Institute of Technology, Kharagpur

In spite of the fact that the Pigeon Pea, *Cajanus,* is widely cultivated in India and Africa, critical studies on its origin and cytogenetics have been largely neglected. After the great work of Vavilov (1926) a beginning was made on the study of this crop in Poona under the leadership of Dr L.S. Kumar. In this paper, I will make a review of the distribution, ancient history and philology, morphology, and cyto-genetics of *Cajanus* and its allied genus *Atylosia* with special reference to our con-tribution from Kharagpur, with a view to tracing the centre of origin of the crop and its wild progenitor.

Distribution of *Cajanus*

The *Index Kewensis* lists about thirty-two species of *Cajanus* distributed all over the tropics; however, only a few are likely to be valid species. Indeed de Candolle (1886) considered the whole complex as one valid species. *Cajanus cajan* Mills. (*Cajanus indicus* Spreng) is found both as wild and cultivated forms. It is a very hardy plant and can grow and complete its life cycle successfully, even in difficult environmental conditions. Hence the wild varieties of the species reported by various botanists are possibly derived from neglected cultivation or casual dispersion.

On the basis of common names, de Candolle (1886) contended that the crop was introduced to the West Indies from the coast of Africa by the slave trade. This is also the opinion of the number of authors on American floras. It was also taken to Brazil, Guiana and into all the warm parts of the American continent. De Candolle quoted Seemann's statement that missionaries introduced it to the Fiji Islands. On the doubtful argument that 'writers of the flora of continental India have only seen the plant cultivated and none, to my knowledge, affirms that it exists in the wild' and that wild varieties had been found in Africa, de Candolle concluded that Africa was the centre of origin. Zukovsky (1962), without giving reasons, favoured an African origin. However, the genus does not show the expected diversity in Africa and Engler (1915) found no evidence in favour of its African origin.

From his vast collection Vavilov (1926) observed considerable variability of this plant in India and concluded that *Cajanus* originated in the Hindustan centre. Shaw, Rehman and Singh (1933) isolated 107 types from samples collected in various parts of India, while Mahta and Dave (1931) isolated 36 types from Madhya Pradesh alone. This is indicative of the varietal wealth of the species in India. Burkill (1953) contended that *Cajanus* was ennobled in India. On the basis of Sanskrit names of *Cajanus* in western Malaya, he believed that it was introduced there from India. This probably happened in the last centuries B.C. when Indian traders discovered how to cross the Bay of Bengal and began to make trading voyages to what was called *Subarnabhumi* or the coast of gold. Long after western Malaya was hinduised, the crop was carried forward to Canton in China in the sixth century A.D., probably during the short period A.D. 502–57. It is not unlikely that through the islands of Indonesia it reached Australia. Relatively recent introductions by British colonists have added to the present day diversity of the plant in Australia. Ecklow and Zeyher (1836) did not mention this crop in their work on the African and Australian flora.

Murdock (1959) in his book *Africa, its people and their culture and history,* has described how many cultivated crops were interchanged between ancient Azania and India. In Mediterranean antiquity, Azania refers to the coast of Kenya, Somalia and Tanzania from Kisimayu to Kilwa, with the islands of Pemba and Zanzibar. He contended that the interchange of crop plants took place both across the Arabian desert by way of the Sabaean Lane, and across the Arabian Sea. Merchant vessels travelling with the monsoon in either direction loaded up at the port of origin with supplies available there and disposed of any surplus at their destination, with the result that the coasts of Azania and western India came early to share the same roster of food plants. Murdock believed that *Cajanus* was brought to eastern Africa in this way in the fifth century B.C. From there the crop moved west to the Congo area and north to the Nile valley. According to Prain and Burkill (1938) the transport of cultivated crops from India to the east African coast is rather recent. They point out that the African end of the Sabaean Lane is inhospitable to most plants, being far too dry for their thrift. Thus the introduction of these crops to Africa from the east was delayed until the Yemenite Arabs had so developed their colonies in Zanzibar and on the Mazunga coast of Madagascar as to make a home for them. This they did between the eighth and eleventh century A.D. Later from the eastern coast the crop moved to equatorial Africa and the Congo area. The Congo area has been the main source from which the crop spread further west, as is evident from such names as Congo pea in English, *pois de Congo* in French, *Ervilha do Congo* in Angola. In Zaire it is known as *kakunda bakishi, Mbaazi* or *Nkol.* It is interesting also that in French the crop is also known as *pois d'Angola.*

Ancient history and nomenclature

There exists no reference to the crop in Vedic literature belonging to the period 1200–600 B.C. Apparently there is also no reference to *Cajanus* in *Ramayana* and *Mahabharata* which date back to 600 A.D. or the end of the Gupta period in India. There is no archaeological evidence so far on this pulse crop.

The earliest reference to the crop in the literature is in the text entitled *Gathasaptasati* written in Maharastrian Prakrit about the third and fourth century A.D. In this text it is referred to as *Tuvari.* The name *Tuvarica* occurs also in *Amarkosa,* a Sanskrit lexicon dated around the sixth century A.D. There are also some synonyms of *Tuvarica, Adhaki* for example. *Adhaki* is also mentioned in *Susrutasamhita* the Sanskrit medical text dated around the sixth century A.D. Piddington (quoted by de Candolle) gave a Sanskrit name *Arhuku* which was not known to Roxburgh. There seems to be little doubt that *Cajanus* was a well known cultivated crop in India by the sixth century A.D. at the latest. These references are drawn chiefly from north Indian and Deccan literature. Whether in the corpus of early south Indian literature, there is any mention of this crop is yet to be determined. There exists no Semitic name for the crop.

The generic name *Cajanus* is derived from the Malaysian name of this plant, *Katjang* or *Katschang*. In the immediate neighbour country, Burma, it is known as *pay-in-chong, pai-si-gong* or *pesigon,* bearing no linguistic relationship with *Katjang.* The local names of the crop in India appear to be derived from two unrelated Sanskrit names, *Tuvarica* and *Adhaki.* The present-day local names can be divided into two major groups – *Tuvaray, Tuvara, Tura, Tur,* etc. in one group and *Arahar, Arhar, Ihora, Oror,* etc. in another. Superficially, the local names in south India seem to be derived from the Sanskrit name *Tuvarica* and those in the north from *Adhaki.* Dr Suniti Kumar Chatterji, the foremost linguist in India (personal communication) points out that neither of the two Sanskrit names are true Sanskrit words with meaningful derivations. He believes that *Tuvarica* was adopted from the pre-existing Dravidian local names like *Tuvarai* or *Tuvari.*

The varieties of the crop cultivated in India can be grossly divided into two chief types – 'Tur' and 'Arhar', with numerous intermediate forms. De Candolle and later Krauss (1927) and Shaw *et al.* (1933) recognised these types as *C. flavus* (Tur) and *C. bicolor* (Arhar). *C. flavus* is a relatively short plant with the standard or upper petal coloured plain yellow and is chiefly cultivated in the Deccan and south India. *C. bicolor* is tall with the standard veined with purple on a yellow background and is cultivated mostly in northern India. From the Sanskrit names akin to the two types, it can be inferred that both types were developed very early in their ennoblement. Whether they developed independently cannot be ascertained.

The relations of *Cajanus* with *Atylosia*

A search for the wild prototype of *Cajanus* led Deodikar and Thakar (1956) to conclude that certain erect species of *Atylosia* are the species closest to *Cajanus.* The two genera belong to the subtribe Cajanae under the tribe Phaseolae. *Atylosia* has a large grooved aril on the seed, and it is separated on this character from *Cajanus,* which has no aril. It must be pointed out that presence of aril is not a strict qualitative character, since in many cultivated varieties of *Cajanus* and in many species of *Atylosia* various degrees of aril development can be seen. Except for size and vigour, certain wild forms of the cultivated crop are virtually indistinguishable from certain erect species of *Atylosia.* The erect species are *A. lineata* W., *A. sericea* Benth., *A. candollei* W. and *A. geminiflora* Dalz. It may be recorded here that in the western Ghats of India the local peasants and tribal people know *A. lineata* and *A. sericea* as *Ban Tur* or Wild Tur. Similarly *A. scarabaeoides* is known as *Ban Arhar* or wild Arhar in Bengal. *A. cajanifolia* is known as *Ban Arhar* in Orissa. Its seeds are eaten, chiefly by the children.

The chromosome number of *Cajanus* is n = 11 (Roy, 1933). Tschechow and Karataschowa (1932) reported the chromosome number of *Atylosia barbata* as 2n = 22. Deodikar and Thakar (1956) carried a detailed study of the chromosomes of *A. lineata* and *A. sericea* and reported the number in both the species as n = 11. From a comparison on the somatic karyotypes they concluded that *Cajanus* has six pairs similar to those in *A. sericea* and four pairs similar to those in *A. lineata.* Kumar,

PLATE 1
(a) *Cajanus cajan* (*C. flavus*, race T-21)
(b) *Atylosia lineata.*
(c) *Atylosia sericea.*
(d) *Atylosia scarabaeoides.*

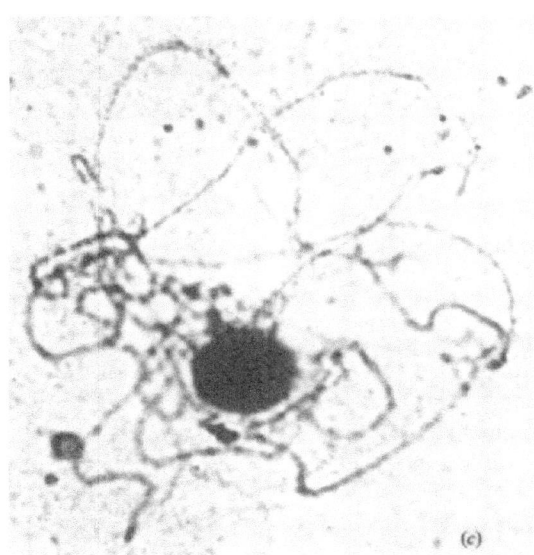

PLATE 2

(a) F$_1$ of *C. cajan* × *A. sericea.*
(b) Habit of *C. cajan* × *A. lineata* hybrid.
(c) Pachytene chromosomes of F$_1$ of *C. cajan* × *A. lineata.* Note almost perfect
pairing of homoeologues.

Table 1   *Success of crossing between* Cajanus *and* Atylosia *spp.*

Cross	Per cent success
C. cajan × *A. lineata*	32.9
A. lineata × *C. cajan*	0
C. cajan × *A. sericea*	9.5
A. sericea × *C. cajan*	2.0
C. cajan × *A. scarabaeoides*	0.2
A. scarabaeoides × *C. cajan*	0

Thombre and D'Cruz (1958) basically confirmed the karyotype of *A. lineata* presented by previous workers, but pointed out that this species, like *Cajanus*, has one pair of satellited chromosomes. Sikdar and De (1967) confirmed the results of Kumar *et al.* (1958) about the presence of a secondary constriction in the short arm of the largest chromosome of *A. lineata* and in the number of median chromosomes. Sikdar and De (1967) and later Reddy (unpublished) worked on the karyotype of *A. scarabaeoides* and reported the number as n = 11. The chromosomes of this species are very much like those of the other two species.

The possibility of hybridisation between *Atylosia* and *Cajanus* was first pointed out by Deodikar and Thakar (1956). Kumar *et al.* (1958) succeeded in obtaining hybrid plants from *Cajanus* × *A. lineata.* The hybrid was partially fertile, with a degree of meiotic irregularity and pollen sterility. At Kharagpur we (Reddy, unpublished) have been able to hybridise *Cajanus* (T-21) with all the three species of *Atylosia* mentioned above (Plates 1*a–d*, 2*a, b*). The percentages of the successful crosses are given in Table 1. In all cases success was greater when *C. cajan* was used as the female parent. The F$_1$s of *C. cajan* × *A. lineata* and *C. cajan* × *A. sericea* were similar in respect of the chief characters in which *Cajanus* and *Atylosia* differ. They behaved as follows:

Atylosia character dominant:
        Pigmented standard petal
        Hairy pods
        Presence of aril on seed
        Shattering pods
        Persistent stipules
        Persistent petals
Cajanus character dominant:
        Lanceolate cotyledonary leaf
F$_1$ intermediate:
        Size and shape of leaf
        Period of flowering
        Size of flower, pod and seed

The pod colour of *Cajanus* is green with purple streaks and of *A. sericea* and *A. lineata* is green. In the hybrids the pods are always uniformly reddish brown. The average number of seeds per pod in the hybrids is close to that of *Atylosia*. Data on pollen fertility and seed setting are given in Table 2.

Table 2 *Fertility of parents and F₁s in* Cajanus *and* Atylosia

Name of plant	Pollen fertility percentage	Seed setting percentage
Cajanus (T-21)	97	99
A. lineata	97	100
A. sericea	96	99
Cajanus × *A. lineata*	78	74
Cajanus × *A. sericea*	50	72

Kumar *et al.* studied the meiosis of the *Cajanus* × *A. lineata* hybrid and reported almost normal behaviour of the chromosomes. Our preliminary work (Reddy, unpublished) on the pachytene chromosomes of the hybrid shows almost complete pairing of all the chromosomes. Only a few terminal and interstitial regions of the homologues occasionally fail to pair. Later stages of meiosis exhibit no abnormality. This indicates a very high degree of homology between the species. However the *Cajanus* × *A. sericea* hybrid presents a different picture. Although pachytene chromosomes show almost normal pairing, a number of abnormalities are met with at later stages with the formation of univalents. Since a minimum of seven bivalents was found in all cases, it is concluded that seven pairs of chromosomes are common to both the genera.

On the basis of habit, morphology, karyotype, hybridisation studies and meiosis *A. lineata* is closest to *Cajanus.* However, before the wild progenitor of *Cajanus* can be precisely determined other erect species of *Atylosia,* e.g. *A. geminiflora, A. candollei, A. cajanifolia* should be thoroughly investigated. Workers on *Atylosia* and *Cajanus* have repeatedly pointed out that it is improper to separate the two genera only on the character of the aril and they should be merged together following the rules of botanical nomenclature. *Cajanus* should be considered as a single polymorphic species.

Distribution of *Atylosia*

Since we accept that *Cajanus* originated from certain species of *Atylosia* the pattern of distribution of the latter may enable us to determine the centre of origin of *Cajanus. Atylosia* is distributed in tropical Asia, Australia and Mauritius. Only two species are reported from the whole of Africa where the genus is rare (Hutchinson and Dalziel, 1927; Thonner, 1915; Hiern, 1896). Indochina has three species (Lecomte, 1910). The *Index Kewensis* records six species in Australia and only two in Mauritius. On the other hand, out of twenty-five known species in the genus, sixteen are widely distributed in India.

Atylosia is abundant in hilly regions of India and is widely distributed over the Western and Eastern Ghats (Fig. 1). It is abundantly found in (1) Nilgiri Hills, (2) Annamalai Hills, (3) Paulghat Hills, (4) Amleoti Ghats, Sawantwadi, (5) Karimalai, and (6) Veligonda Hills. The greatest differentation and greatest abundance of individuals of *Cajanus* and *Atylosia* is found in the broad leaf evergreen forest areas of the Western Ghats and the Malabar coast, and it can be concluded that these constitute the centre of origin of *Cajanus.* From this centre the whole Indian sub-continent has become the area of diversity. No comparable secondary centre exists.

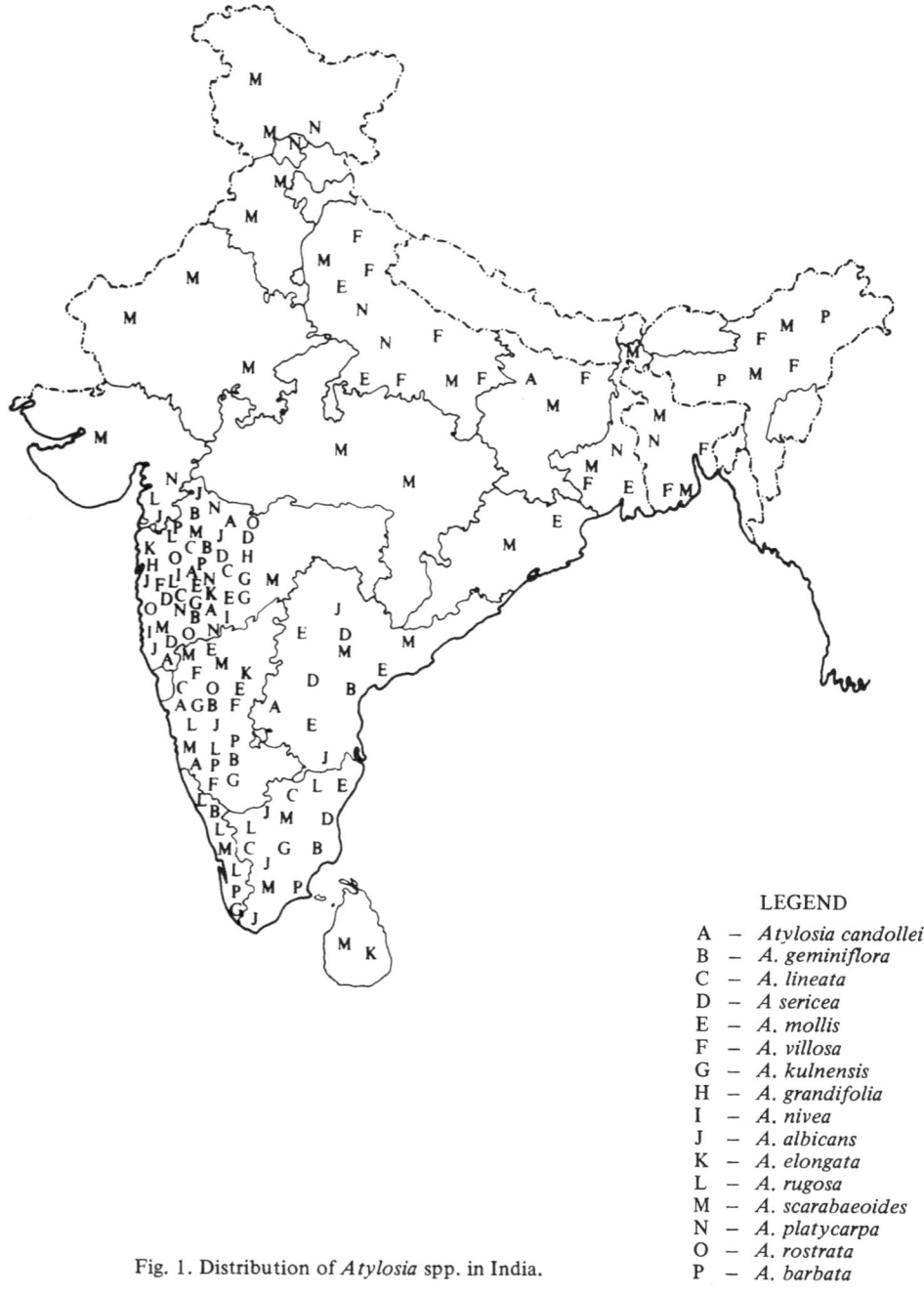

Fig. 1. Distribution of *Atylosia* spp. in India.

LEGEND

A – *Atylosia candollei*
B – *A. geminiflora*
C – *A. lineata*
D – *A sericea*
E – *A. mollis*
F – *A. villosa*
G – *A. kulnensis*
H – *A. grandifolia*
I – *A. nivea*
J – *A. albicans*
K – *A. elongata*
L – *A. rugosa*
M – *A. scarabaeoides*
N – *A. platycarpa*
O – *A. rostrata*
P – *A. barbata*

Concluding remarks

Cajanus and *Atylosia* have the same chromosome number, n = 11, and this is the basic chromosome number. Thus the development of the crop took place without any alteration in the chromosome number. It is likely that *Cajanus* originated through selection of gene mutations with a single species. The characters chiefly concerned have been those that have been very generally selected in crop plant domestication, and include size and vigour of the plant, and especially the size and non-shattering character of the pod, and size and number of seeds. Yet the crop does not show the great variability in pod and seed characters found in the most advanced cultivated plants and it may be concluded that the ennoblement of *Cajanus* is comparatively recent, a fact which is corroborated by historical accounts.

Acknowledgements

The author is most grateful to Mr L. Janardhana Reddy for permitting him to use his unpublished results and for collecting certain information. He is also grateful to Dr Suniti Kumar Chatterjee, National Professor in Linguistics, for a fruitful discussion of the derivation of various names of *Cajanus* and to Mr Dilip K. Chakrabarti, Lecturer in Archaeology, Calcutta University for information on the crop in ancient Indian literature. Thanks are also due to Dr A.K. Chatterjee of our Department, for extensive discussion and for permitting the author to use his personal library.

Cotton

V. SANTHANAM
I.A.R.I. Regional Station, Coimbatore and

J.B. HUTCHINSON
St John's College, Cambridge

Origin and distribution

The genus *Gossypium,* to which the cottons belong, comprises approximately thirty tropical and subtropical species that have been assigned to six cytologically defined diploid genome groups, A, B, C, D, E and F and one tetraploid genome group 'AD' (Phillips, 1966; Phillips and Strickland, 1966). Extensive cytogenetic data on their inter-relationships have been reported primarily by Skovsted (1934, 1937), Webber (1939), Beasley (1940, 1942), Douwes (1951) and Phillips (1963).

The lint-bearing species of *Gossypium,* which are the true cottons, are four, the diploid *G. arboreum* and *G. herbaceum* indigenous in Asia and Africa, and the tetraploid *G. hirsutum* and *G. barbadense* indigenous in the New World. It has been shown that the New World cottons are natural amphidiploids combining the A genome from a taxon of the Asiatic diploid group and a D genome from a taxon of the American diploid group. The dating of the origin of the cottons has been the subject of two major theories, the ancient origin theory of Harland (1939) and Stebbins (1947) and the recent origin theory of Hutchinson, Silow and Stephens (1947). The theory that the cottons are of recent origin was put forward by Hutchinson and his colleagues to account for the fact that wherever true cottons are found growing wild, their closest relatives are the earliest known cultivars in the same region, and not wild cottons growing at a centre of origin. They argued that such a situation would arise if the cottons now growing wild were escapes from cultivation, and not truly wild forms. If this were so, it followed that the very large genetic gulf between the cottons and the wild species of *Gossypium* arose as a result of the selection by man of mutant types which bore spinnable lint.

As knowledge of the distribution and relationships of the primitive cottons increased, the theory that they are of recent origin became progressively more difficult to sustain. Hutchinson (1954) accepted *G. herbaceum* race *africanum* of southern Africa as a truly wild form. He was then led to postulate spread from southern Africa to Arabia and the northwestern region of the Indian subcontinent following the adoption by Man of *africanum* or something resembling it, for textile purposes. He supposed that *G. arboreum* had differentiated from *G. herbaceum* under cultivation in northwestern India and Pakistan.

Elucidation of the time scale of agricultural development generated doubts of the truth of this hypothesis. These were further increased by Chowdhury and Buth's (1971) demonstration that, about 2500 B.C., seed of a true cotton, albeit with a primitive lint, was used as feed for livestock by Neolithic people in the Nile Valley who had not learned the textile craft.

There remained the problem of the almost universal close relationship between wild-growing cottons and primitive cultivars in the same area, and this was resolved by Hutchinson (1970). He showed the relevance of Thoday's (1964) work on disruptive selection in *Drosophila* to the genetics of a plant selected for domestication, and grown in the same area as its wild relative. Intercrossing leads to the exchange of genes that are selectively neutral, while genes selected by man will increase in the cultivar, and those favouring survival in the wild will persist in the wild type. Moreover, the wild type will develop an association with the cultivar, will spread with it, and will persist on field margins and in abandoned clearings. Thus the association between the wild and cultivated types, both in morphology and in distribution, is accounted for by the genetic circumstances of domestication and spread. This is an amplification of the earlier hypothesis, in that the types now found growing wild are not escapes from cultivation but are associates of the cultivars. They have been dominated both in genotype and in distribution by their cultivated relatives.

The origin of the true cottons may now be reconsidered. At about 2500 B.C. cotton seed was being fed to livestock in Nubia by people who did not practise the textile art. The earliest civilisation known to have spun and woven cotton was the Harappan in the Indus Valley in Pakistan (2300–1750 B.C.), and for many centuries the cotton plant was known outside India only in travellers' tales. The cotton textiles of the Harappan civilisation were the product of a sophisticated textile craft, and cannot be regarded as the earliest in the Indian region. Thus at the earliest agricultural levels yet discovered in the Nile Valley in Africa, and in the Indus Valley on the Indian subcontinent, true cottons were already present. Moreover, wild and weedy types are to be found associated with primitive cultivated types in both Old World species. In *G. herbaceum* they are known from the coastal strip northwest of Karachi, through northern Baluchistan to South Yemen, Ethiopia and Sudan, and even to west Africa south of the Sahara. In *G. arboreum* they have been recorded by Watt (1907) from Kathiawar, Gujarat, Khandesh, and the Deccan. Thus there remains nothing to support Hutchinson's suggestion that *G. arboreum* was derived from *G. herbaceum* under domestication, and every reason to accept the two species as having been separately adopted from the wild.

In *G. herbaceum*, race *africanum* is isolated in southern Africa. On this present interpretation there is no reason to postulate that it contributed to the cultivated cottons. It may be supposed that the Arabian and Baluchistan race *acerifolium* was growing wild in that region, and was adopted by civilised communities in the area. In *G. arboreum* there is nothing to indicate whether the species was already subdivided when it was first domesticated. It may be, however, that the differentiation of three perennial races, *burmanicum* of northeastern India, *indicum* of western India and the peninsula, and *soudanense* of northern Africa, antidated domestication, and that each contributed separately to the cultivated cottons. However, it seems likely that it was in Gujarat or Sind that *G. arboreum* cottons were first brought into cultivation (Hutchinson, 1971*b*).

Evidence on the origin of the New World cottons will not be discussed in this paper, which is concerned with evolution in the Indian region. It is sufficient to say that it must be accepted that *G. barbadense* and *G. hirsutum* existed as distinct species in the wild, and their cultivated derivatives were separately domesticated.

It follows that the extent of evolutionary change under domestication has been much less than was postulated by Hutchinson, *et al.* (1947). Change in response to the demands of cultivation has been both substantial and rapid, but it has gone on within the species, and has not led to the emergence of new species as they supposed.

Indigenous cottons in India

The history of the cotton crop in India is one of change in response to changing agricultural, economic and political forces. The establishment of the cotton crop as the raw material of India's major textile industry depended on the adoption of perennial cottons, and their improvement in yield and in quality. The fine spinning and weaving commonly associated with Dacca muslins, but practised widely in India (Hutchinson and Ghose, 1937) was made possible by the selection of fine quality types among the perennials. These cottons were all forms of *G. arboreum.* A review of Watt's (1907) account (Hutchinson, 1959) shows that the first record of *G. herbaceum* in India is Hove's, in 1787. Moreover, it appears that this intro-duction, probably from what is now Iran, was stimulated by a developing interest in the potential of growing cotton as an annual.

This marks the beginning of the second phase in the development of the Indian cotton crop. Annual types were selected by taking advantage of the morphology of branching of the cottons, and selecting for a switch from vegetative to fruiting branches at an early node on the main stem. This was successfully achieved in the two perennial races of *G. arboreum, indicum* and *burmanicum,* and gave rise to the annual *indicums* of peninsular India and the annual *bengalenses* of northern India. The former were on the whole of good quality and with a low ginning out-turn, similar to the old quality perennials that they succeeded. The latter included types with coarse staple and high ginning out-turn, and these were to be of particular importance in the next stage of development. In *G. herbaceum* also, further progress was achieved in the development of the annual habit. In Gujarat, types were grown which took a long period to produce a crop, and in which one of the characters associated with the perennial habit persisted. Though fruiting branches were developed early on the main stem, the flower buds they carried were shed in wet weather. Only with the onset of the dry season did the flower buds complete their development and produce flowers and bolls. This must be regarded as a primitive character, and in the earliest fruiting *herbaceums* that have been developed, both in Gujarat and in peninsular India, it has been bred out, and fruiting is morphologically determined instead of being physiologically limited.

Up to this stage the Indian cotton crop was grown for Indian use. The impact of British trading interests was disastrous for the Indian textile industry, and it collapsed under the competition from machine-made piece goods imported from Lancashire. The only compensation for the loss of the local market was the devel-opment of exports of raw cotton. There followed a long period when much of India's crop went to the Far East. The development of the railway system opened up areas previously isolated, and the raw cotton trade became integrated nationally. In northern India the coarse, higher-ginning forms of *bengalense* were favoured, since yield of seed cotton and ginning out-turn were of prime importance to the grower. These types spread southwards and invaded the *indicum* areas of Maharashtra and

Andhra Pradesh, and the *herbaceum* and *indicum* areas of Gujarat. The quality cottons were mixed with the new coarse types, and since there were no safeguards against adulteration, the cultivator in self-defence grew the high-ginning coarse cottons in place of the low-ginning types, for which he could no longer be sure of receiving a premium.

These trends were matters of some concern to the administration. Hutchinson and Ghose (1937) have given an account of the interest that was evident in the 1840s in the improvement of the cottons of Malwa, and have drawn attention to the *Return : Cotton (India)* (1847) in which the work done for the improvement of cotton in India as a whole is reported. Interest centred at that time on the possibility of replacing *desi* (Asiatic) cottons with New Orleans (*G. hirsutum*) cottons. In the *Return* (1847) is given an account of the work of two American cotton planters, brought in by the East India Company to demonstrate the cultivation of the New Orleans seed. In the diary of one of them reproduced there, is a graphic account of the progress of the crop, from good germination and seedling development, through a phase of increasing unhealthiness, to a final complete collapse under what to a modern reader was clearly a devastating jassid attack. There were later attempts to introduce Upland Georgian cottons, particularly at the time of the cotton famine that resulted from the American Civil War, but these also were abortive, and for the next half-century the Indian crop was grown virtually entirely from *G. arboreum* and *G. herbaceum.*

The introduction of New World cottons

The significance of the early and unsuccessful introductions of New Orleans and Upland Georgian types of *G. hirsutum* only appeared later in the evolution of the Indian cotton crop. In recent times it has become very great, and for an understanding of the genetic factors contributing to the emergence of *G. hirsutum* cottons as a major component of the Indian crop, a brief account is necessary of the differentiation of this species, most of which took place outside India. Hutchinson (1951) has described seven races of *G. hirsutum,* of which only one, race *latifolium,* has been of importance in the development of the modern cottons. Race *latifolium* is annual in habit, and when it was introduced from Mexico to the southern United States, it gave rise to the commercial cottons that have become known as Uplands.

Distinct races of Upland have arisen in other parts of the world from Upland stocks introduced from the United States Cotton Belt. The Indian race derived from the New Orleans and Upland Georgian introductions, and modified in response to Indian conditions, was one such. Another was the African Upland derived from high-quality American varieties introduced to British African territories during the first and second decades of the twentieth century and distinguished by the development of its own characteristics under local selection. Rather later in the twentieth century, Uplands introduced into central Asia were developed to supply Russian needs for cotton, and there also a distinct stock has arisen. The Uplands have in common the history of passage through the United States Cotton Belt, where short-day fruiting was eliminated, and only long-day tolerant types survived. Other *latifolium* cottons were taken direct from Mexico to the Philippines and to Cambodia, and these formed a southeast Asian stock that had not been subject to selection for long-day tolerance. It later became of importance in south India under the name 'Cambodia'.

For convenience of reference in discussing the way in which *hirsutum* cottons have been bred in India, these distinct commercial stocks of race *latifolium* will be named as follows:

(a) U.S. Upland: varieties from the United States Cotton Belt.
(b) African Upland: varieties bred in Africa from Upland stocks originally from the U.S.A.
(c) Russian Upland: varieties bred in Russia from Upland stocks originally from the U.S.A.
(d) Indian Upland: varieties bred in India from stocks derived from the New Orleans and Upland Georgian introductions.
(e) Cambodia: varieties bred in India from the *latifolium* stock established in southeast Asia.

Many of the most recent *hirsutum* varieties bred in India are derived from crosses between these stocks, and selected for suitability to Indian conditions.

The next stage in the development of the Indian crop began about the beginning of the twentieth century. British and Indian capitalists had established modern cotton mills in Indian cities, and as the cotton industry developed, a demand arose for a range of qualities of raw cotton, including an increasing amount above the base level of the coarse, high ginning *bengalense*. In northern India Burt and others (Burt, 1913) reselected Upland strains from the survivors of the early introduction, still persisting in small numbers in crops of *desi* cotton. Over the half-century of survival under the partial shelter of the jassid-resistant *desis,* they had been selected for the leaf and stem hairiness that gives resistance to jassid attack. Whereas their ancestors had failed miserably, these selected descendents cropped satisfactorily under suitable conditions, and with the extension of irrigation from the rivers of the Punjab particularly, a substantial crop of Indian Upland cotton was established.

In peninsular India also, some crops of New Orleans or Upland Georgian origin such as Dharwar American, were established in a similar way, but much more important was the introduction in 1904–5 of the cotton that became known as Cambodia. This cotton was first introduced into the Madras Presidency from Pondicherry through two independent channels, in 1904 by Benson and in 1905 by Steel (Sampson, 1911), and on this the crop was established. It proved very successful under irrigation in parts of Madras, and was tried extensively in other parts of India. Cambodias made some contribution in Malwa, where they were introduced by Coventry, but in general they did not compete successfully with Upland types north of Madras State. Attempts to grow them in the Punjab and what is now Pakistan came to nothing because they fruited so late in northern regions.

Cotton improvement

By the end of the First World War the demands of the Indian cotton industry for reliable supplies of cotton of known and repeatable quality were such that a move was made to establish quality controls and crop improvement schemes. The basis of quality control was the Cotton Transport Act and the system of appeals and arbitrations organised by the East India Cotton Association. Crop improvement was undertaken by setting up the Indian Central Cotton Committee, which organised quality studies at its own Technological Laboratory, breeding work on Government Stations and at the Institute of Plant Industry, Indore, and pedigree seed supply schemes. Under the Cotton Transport Act it became possible to protect zones producing quality cotton from adulteration with coarse cotton brought in from outside, and with this security and the establishment of quality seed stocks, the characteristic quality areas of the Indian crop reappeared. In particular in the areas of middle and western India where the encroachment of coarse *bengalense* had been most serious, old quality varieties reappeared, and the position was consolidated and improved by the issue of a series of new types of hybrid origin, in which yield and quality had been successfully combined.

The middle 1930s was a period of great interest in, and debate on, plant breeding policy, and at the Indian Central Cotton Committee's first Cotton Conference Ramanathan, working at Coimbatore (Ramanatha Ayyar, 1937), put forward the proposition that cotton breeding should be concentrated on American-type cottons at the expense of the *desis*. Hutchinson, from Indore, argued that the potential of the *desis* had not yet been exploited, whereas the acclimatised American types had probably reached the limit of successful cultivation. Ramanathan and Hutchinson were influenced by their experience on their respective stations, and the argument was inconclusive. Nevertheless, the debate in 1937 foreshadowed the choices that lay before those who faced the next and latest phase in the evolution of the Indian crop.

Two factors determined the onset of this latest phase, the partition of 1947 which cut off from India the greater part of the Indian Upland crop of the Punjab, and the development of modern insecticides, which made effective control of cotton pests possible. The Indian mills, deprived of the Indian Upland cottons of the Punjab and faced with the trend of demand towards better quality textiles, demanded of Indian growers a switch to longer and finer stapled varieties. The only way that this could be done in the short term was by the replacement of *arboreum* cottons by *hirsutums*. And with the protection against insect pests afforded by the new range of insecticides, it became possible economically to introduce Indian Upland varieties in many areas where only *arboreums* could be grown before. Thus in large measure Ramanathan's policy became both possible and remunerative, and the ground for Hutchinson's doubts was removed by innovations in pest control.

The present state of cultivation of the three species of cotton grown in India may now be summarised. Cotton belonging to *G. arboreum* races *bengalense* and *indicum* accounts for 25 per cent of the total cotton production and 29 per cent of the area devoted to the crop. The southward spread of *bengalense* which caused extensive deterioration in the quality of the crop, affected particularly parts of Madhya Pradesh (Malwa and Nimar), Gujarat (the Mathio area), and Maharashtra (Khandesh and Vidarbha). In these areas good-quality selections from the local mixtures were distributed, followed by a series of lines derived from *bengalense* X indicum crosses. These combined better quality with good yield and higher ginning out-turn, and in the latter stocks, wilt resistance also.

Further south in Maharashtra the Marathwada tract was known for the production of the finest quality *arboreum indicum* cottons, and once the risk of adulteration from outside had been removed, the major selection pressure was for the development of superior spinning lines of the indigenous Gaorani. Good quality *indicum* cottons are also grown in Andhra Pradesh (the Northerns and Hingari of Rayalaseema and the Coconadas of the coastal zone), and in Madras (the Karunganni tract). Improvement has come in these areas from selection in *indicums* and their hybrids, and by the elimination of Uppam, a coarse *herbaceum* component of Karunganni.

Cotton belonging to *G. herbaceum* accounts for 24 per cent of the total cotton production and 21 per cent of the area devoted to the crop. *G. herbaceum* is confined to western India, from Kutch in Gujarat in the north to Dharwar in Mysore in the south. In northern Gujarat the distinctive Wagad type is grown, with round bolls which hardly open when ripe. Of the more usual open-bolled type, the Broach and Surti cottons were characteristically long-duration forms, and coarse-stapled, high-ginning types had spread in the period before 1920. Further south, short-duration types were grown. Up to the 1930s, good quality *herbaceums* were grown only on limited areas in southern Gujarat. Since then, selection in local stocks and in hybrid material has given rise to a range of types of better quality and shorter duration, and to the establishment of a high level of wilt resistance. Thus the general level of quality of the *herbaceum* crop has been raised.

The establishment in India of *hirsutum* cottons of two different origins has been described above. Improvement in the Cambodia and the Upland lines began separately, but increasingly the two have been combined in hybrid stocks to form the basis of recent breeding work. The first improved strain of Cambodia which gained popularity, Co2, was issued in Madras in 1929. Co2 has been extensively used in later work. The next significant step in the Cambodia tract was the issue of Co4, an earlier, better quality strain from a hybrid between Co2 and an African Upland. This was early enough to allow of double cropping. Further improvement by selection led to the issue of a better quality, bacterial blight resistant stock, MCU-1.

The special feature of the strains Co4 and MCU-1 was that they were suitable for cultivation both in the winter season, i.e. August–September to February–March, in the Coimbatore tract, and the summer season, i.e. March to August in the Rajapalayam tract. The extension of cotton cultivation in the summer season in the southern districts of the state, under the relatively long day conditions, was made possible only after hybridising Cambodia Co2 with African Upland cottons, which are tolerant of long days.

Further improvement was gained by selection in material derived from crosses involving African Uplands from South Africa and Uganda, Acala and Sealand (Uplands) from the United States, and a remote derivative from Sea Island (*G. barbadense*). MCU-4 was released in 1967 for the summer Cambodia cotton tracts, and MCU-5 in 1968 for the winter tracts. These two cottons can be used up to 60s counts in Indian mill conditions, and they have placed Tamil Nadu in the forefront in the production of quality cottons.

Much of the early work on the Indian Uplands was carried on in what is now Pakistan, where 4F was released in 1914, and LSS in 1933. After partition in 1947, efforts were made to develop Upland cotton cultivation in India, in east Punjab and

Haryana, and in neighbouring areas of Rajasthan and Uttar Pradesh. The first new strain issued was an early maturing selection from a Sind variety. Earliness has been a major object of selection, partly as a means of evading pink bollworm attack. Close attention has also been given to jassid resistance, and resistance to bacterial blight is also important. Improvement in quality has been achieved and further advances are in prospect, but this region is likely to remain primarily a producer of medium staple in the Upland range.

The early attempts made to introduce *hirsutum* cottons in the Gujarat tract met with failure. This led to an attempt to combine the fibre qualities of *hirsutum* cottons with the general adaptability of Asiatic cottons *arboreum* and *herbaceum*. Interspecific hybridisation experiments were started in Gujarat in the second quarter of the present century (Desai, 1927). The crosses are difficult to make, and the F_1s, when obtained, are highly sterile.

The cross *hirsutum* (Dharwar–American 2-6-5) × *arboreum* (Gaorani-6) was made in 1936. In 1940, four viable seeds were obtained from an F_1 hybrid which was being perpetuated as a ratoon plant and one of the plants raised from these seeds proved to be a tetraploid. This plant was crossed to Cambodia Co-2 (*G. hirsutum*), and after continuous selection in subsequent generations the variety 170-Co-2 was released in 1952 under the name 'Deviraj'. This variety, which has extra-long staple, has spread widely in cultivation in many parts of Gujarat, Maharashtra and Mysore.

Since then, other Upland-type strains have been released, Devitej with 1027 ALF (*herbaceum*) in its ancestry, and Gujarat 67 with red *arboreum*. The release of these cottons derived from interspecific triploid hybrids constitutes a major genetic advance in the tetraploid cottons. They have spread extensively in middle and southern Gujarat, replacing *herbaceum* cottons.

In Madhya Pradesh and Maharashtra the pressing need for Upland staple following partition led to replacement of the traditional *arboreum* cottons wherever possible. Locally acclimatised Uplands were available in the area. Buri had been a component of the Maharashtra Bani crop for many years. The Cambodia in Madhya Pradesh was the result of mixing and hybridisation between acclimatised Upland and Coventry's introduction of Cambodia. The first improved strains were straight selections from these stocks, Buri 147 being widely distributed. Later, Buri 0934 and Indore 2 were crossed with a derivative of a Cambodia × *tomentosum* cross carrying extra jassid resistance with the *tomentosum* hairiness factor. Current seed issues carry the characteristic dense leaf hair of *tomentosum*. High yields from *hirsutum* cottons depend in these areas on supplementary irrigation, but with improved adaptability and a quality that commands a premium, they have spread considerably on rain-fed lands, replacing *arboreum*.

In Dharwar in Mysore State, Upland cotton was established very early. The first improved strain, Gadag 1, was a selection from the acclimatised stock. Gadag 1 has been the basis of further improvement through crosses with Co2 from Coimbatore. The derivative Laxmi has spread widely in neighbouring states as well as in Mysore. Breeding for other Mysore cotton areas has been based on a broad crossing programme, including besides Mysore, Coimbatore and Surat types, strains from Uganda and the United States.

Table 1 *Species composition of cotton grown in India*

Species	Pre-War 1938−9		Partition 1947−8		1955−6		1960−1		1969−70	
	Area*	Per cent	Area*	Per cent	Area*	Per cent	Area*	Per cent	Area*	Per cent
G. hirsutum	1.5	2	1.4	3	16.3	20	22.0	29	38.7	50
G. barbadense	–	–	–	–	–	–	0.04	v.s.	0.01	v.s.
G. arboreum	52.1	64	27.9	65	44.0	54	34.5	45	22.4	29
G. herbaceum	27.1	34	13.9	32	20.5	26	20.3	26	16.0	21
Total	80.7		43.2		80.8		76.8		77.1	

*in lakh (100 000) hectares

Table 2 Improved varieties of New World cotton grown in India, classified according to species, staple length and average mill-spinning capacity

Species	Average mill-spinning capacity (counts)	Staple length classification (32nd inch)					
		Extra-long (38/32 and above)	Superior long (34/32 to 37/32)	Long (31/32 to 33/32)	Superior medium (28/32 to 30/32)	Medium (25/32 to 27/32)	Short (24/32 and below)
G. barbadense	Above 80s	Sujata	–	–	–	–	–
	61s to 80s	S.I. Andrews	–	–	–	–	–
G. hirsutum	41s to 60s	Gujarat-67 MCU-4 MCU-5	Hybrid-4 Varalaxmi	B-1007	Mahalaxmi (1301 DD)	–	–
	31s to 40s	–	MCU-2 Krishna	Bharati MCU-1 B.147 Badnawar-1 Mysore-14 PRS-72 Mysore Vijaya Deviraj Khandwa-1	P.216-F Pramukh A.51-9 (Narbada) Laxmi Khandwa-2 J.34	LSS Hampi	–
	21s to 30s	–	–	–	320-F H.14	C.Indore-1	–

Table 3   Improved varieties of Old World cotton grown in India, classified according to species, staple length and average mill-spinning capacity

Species	Average mill-spinning capacity (counts)	Staple length classification (32nd inch)					
		Extra-long (38/32 and above)	Superior long (34/32 to 37/32)	Long (31/32 to 33/32)	Superior medium (28/32 to 30/32)	Medium (25/32 to 27/32)	Short (24/32 and below)
G. arboreum	31s to 40s	–	–	K-7, K-8	Gaorani-46 Nandicum C.J.73 (Sanjay)	–	–
	21s to 30s	–	–	–	Virnar AK-235 AK-277 Adonicum Y-1	Maljari Coconadas-2 Coconadas-741	–
	11s to 20s	–	–	–	–	–	Ganganagar-1 Shyamali
	10s and below	–	–	–	–	–	231-R G.27
G. herbaceum	31s to 40s	–	–	–	3943 Digvijay Suyodhar	–	–
	21s to 30s	–	–	–	J.797 Jayadhar	Westerns-1	–
	11s to 20s	–	–	–	–	Kalyan	–

The Indian cotton crop is the most diverse in the world, both in its botany and in its quality range. Three of the species contributing to the cotton of commerce are widely grown. The fourth, *G. barbadense,* includes the highest qualities in the cottons, and as far back as 1830 an attempt was made to grow it also in India. This, and subsequent attempts up to 1947, ended in failure. However, the outcome of trials initiated in 1947 was that the variety Andrews Sea Island was identified as promising in 1955—6, but the expansion of the area sown has been limited. More recently, a reselection from Egyptian Karnak has performed well at Coimbatore. It was released in 1969 under the name Sujata, and the upper limit of spinning values of cotton grown in India has been raised thereby to 100s counts. The production of such Indian-grown, high-quality cotton can be stepped up substantially, provided it obtains on the market a premium price commensurate with its quality.

Attempts at producing suitable F_1 hybrids in cotton for commercial cultivation have been made since 1948, especially at centres like Surat and Coimbatore. A hybrid based on Gujarat 67 X American Nectariless named Hybrid 4, released from Surat (Patel, 1971), has spread in commercial cultivation to about 40 000 hectares in 1971.

Conclusion

The species composition of the Indian crop pre-war, at partition in 1947, at the end of the first two Five Year Plans and at the beginning of the Fourth Plan period, is set out in Table 1. In 1938—9 nearly 98 per cent of the area under cotton was sown to the *desi* (Asiatic) species and only 2 per cent to *G. hirsutum.* By 1969—70, the New World cottons accounted for 50 per cent of the area. Better quality has been attained in all species. The quality standards of current varieties are given in Tables 2 and 3.

The change in the quality composition of the Indian cotton crop in response to the needs of the industry following partition has been an achievement of major importance. Yield levels have not been raised as they have been, for example, with wheat, and Indian cotton research workers are now greatly concerned with the prospect of raising yields. High-yielding varieties are available, and perform well where water supply, fertility level and pest control are well managed. New approaches to yield improvement are being studied, as for instance, in Uplands, the possibility of using the 'cluster' character to permit much closer plant spacing and narrow leaf to improve light penetration through the leaf canopy of a high plant population. New plant types may be expected to contribute to the campaign for better yields, but over most of the cotton area the basic need is for better water management, higher fertility levels, and well managed and economical pest control. Thereby will yield potential be realised in field practice.

Okra*

A.B. JOSHI, V.R. GADWAL and M.W. HARDAS
Division of Genetics, I.A.R.I., New Delhi

Introduction

Abelmoschus esculentus (L.) Moench, commonly known as okra or lady's finger (Hindi equivalent, *bhindi*) is an important vegetable crop in the tropical and subtropical countries of the World. Earlier it was named *Hibiscus esculentus* under the section Abelmoschus in the genus *Hibiscus,* established by Linnaeus in 1737. Medikus (cited by Hochreutiner, 1924) proposed the generic name, *Abelmoschus,* and Schumann (1895) also recognised this position. Yet, for many years most botanists continued to hold the opinion that it deserved only a section in *Hibiscus,* and not a separate genus, although they recognised the characters separating it from *Hibiscus.* Hochreutiner (1900) held this view, but later (Hochreutiner, 1924) reinstated the genus *Abelmoschus* on the basis of the constant distinguishing feature, namely caducous calyx. In this genus, the calyx, corolla and the stamens are fused together at the base and fall off as one piece after anthesis. Hochreutiner listed fourteen species under *Abelmoschus*; later many new species were added. The *Index Kewensis* lists about thirty species under *Abelmoschus* in the Old World and four in the New World.

The cultivated okra is of Old World origin. Vavilov (1951) concluded from his phytogeographic studies that it arose as a cultigen in the Abyssinian region. Murdock (1959) suggested that it arose in west Africa early in the history of agriculture. Wild species of *Abelmoschus* are indigenous in Africa, and it will be shown later that some cultivars from the Sudan display primitive characters. Complete elucidation of the evolutionary history of *Abelmoschus esculentus* must depend, therefore, on studies of the African members of the genus. The present study is concerned with the cytogenetic relationships of *A. esculentus* and three wild species native in India, *A. tuberculatus, A. moschatus* (syn. *Hibiscus abelmoschus,* Skovsted, 1935) and *A. ficulneus* (Hardas and Joshi, 1954).

The chromosome number of *A. esculentus* has been variously reported to be $n = 59–72$ (Datta and Naug, 1968; Kuwada, 1966; Roy and Jha, 1958; Skovsted, 1935; Ford, 1938). Joshi and Hardas (1956), after critically examining the chromosome number in a large number of varieties of *A. esculentus,* came to the conclusion that the correct chromosome number of cultivated okra is $n = 65$. Ford (1938) working on Ceylon material, reported $n = c.\ 33$, and Teshima (1933) reported $n = 36$ from a Japanese line. Unfortunately, we were unable to obtain seed of these stocks, either from Ford, or from Professor Kuwada in Japan.

* This paper includes the results reported in a thesis approved for the award of the Ph.D. degree of the Postgraduate School, I.A.R.I., in 1966 to V.R. Gadwal.

Pal, Singh and Swarup (1952) described a new species, *Abelmoschus tuberculatus* Pal *et* Singh which has been collected from several parts of India. Joshi and Hardas (1953) reported n = 29 as the chromosome number for this species. This was a new chromosome number, not hitherto reported in the genus *Abelmoschus*. Joshi and Hardas (1956) studied meiosis in the F_1 hybrid *A. esculentus* (n = 65) × *A. tuberculatus* (n = 29), and observed that the 94 chromosomes in the hybrid almost invariably associated themselves in meiosis as $29_{II} + 36_I$. This suggested that cultivated okra was an amphiploid comprising two genomes: one with 29 chromosomes and the other with 36 chromosomes, the former being homologous with that of *A. tuberculatus*. The question then arose as to which species contributed the 36-chromosome genome to *A. esculentus*. This aspect of the problem is presented and discussed in this paper.

The material studied comprised:

(*a*) A collection of okra cultivars (*A. esculentus*), of which all those examined had n = 65 chromosomes. This genome will be given the symbol E. Since it has not been possible to obtain the stocks of *A. esculentus* on which the n = 36 chromosome numbers were determined, consideration in the present paper will be confined to the n = 65 material.

(*b*) *A. tuberculatus* (Indian) with n = 29 chromosomes (genome T).

(*c*) *A. moschatus* (Indian) with n = 36 chromosomes (genome M).

(*d*) *A. ficulneus* (Indian) with n = 36 chromosomes (genome F).

Genome E (= 65) has been shown to include a component (T') homologous with genome T (= 29). Then the question to be examined is whether the residue of 36 chromosomes of E, which may be designated Y, is homologous with either of the genomes M or F.

An answer to the question is sought through the following studies:

(*a*) meiotic behaviour of diploid and haploid *A. esculentus;* (*b*) meiotic behaviour of *A. esculentus* × *A. tuberculatus;* (*c*) morphological features and meiotic and breeding behaviour of the synthetic amphiploid *tuberculatus–ficulneus;* (*d*) meiotic behaviour of F_1 (*esculentus* × *ficulneus*) and F_1 (*esculentus* × *moschatus*); and (*e*) consideration of variation in fruit morphology in *esculentus* and the aforementioned three species of *Abelmoschus* occurring in India.

Cytological observations

A. esculentus

A meiotic metaphase plate from a pollen mother cell of *A. esculentus* cv. Pusa Makhmali is illustrated in Fig. 1. Meiosis was normal, and 65_{II} can be clearly counted. Two haploid plants were detected by us in advanced generations of intervarietal hybrids of *esculentus*. Meiotic metaphase plates showed sixty-five univalents (Fig. 2), thus confirming n = 65 as the chromosome number of *esculentus*.

A. esculentus × A. tuberculatus

Earlier studies by Joshi and Hardas (1956) showed that the 29 chromosomes genome (T) of *A. tuberculatus* is homologous with 29 chromosomes (subgenome T') of

102

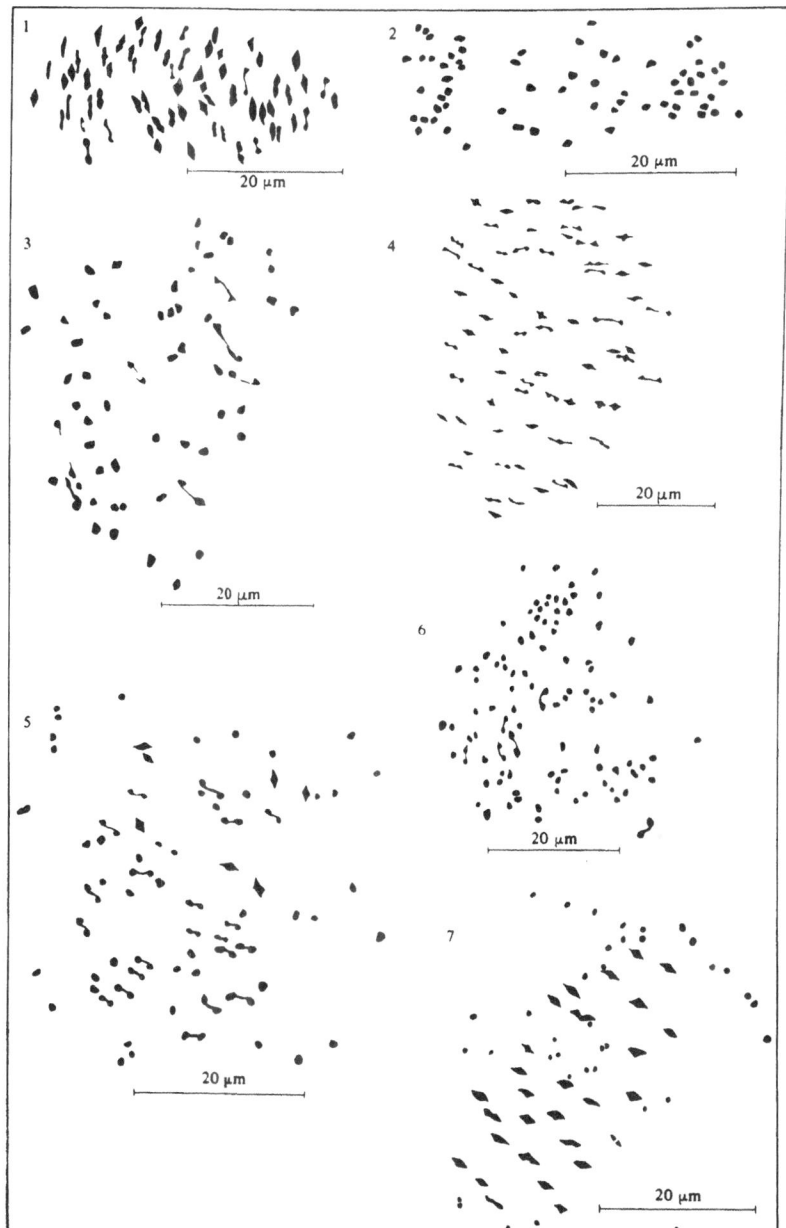

Figs. 1–7. Pollen mother cells at meiotic metaphase I showing:

1   65_{II} in *esculentus* cv. Pusa Makhmali
2   65_I in a haploid of *esculentus*
3   $7_{II}51_I$ in F_1 (*tuberculatus* × *ficulneus*)
4   65_{II} in amph. (*tuberculatus* × *ficulneus*)
5   $28_{II}45_I$ in F_1 (*esculentus* × *ficulneus*)
6   $6_{II}89_I$ in F_1 (*esculentus* × *moschatus*)
7   $29_{II}36_I$ in F_1 (*esculentus* × *tuberculatus*)

A. esculentus. A meiotic metaphase I of this cross is illustrated in Fig. 7, confirming the expected configuration of 29_{II} and 36_{I}.

A.tuberculatus × A. ficulneus

Reciprocal hybrids between these two species are readily obtained. Meiosis was studied both in the F_1 hybrid and in its polyploid derivative obtained by colchicine treatment. The chromosome constitution of the F_1 is T(29) + F(36) = 65. Meiosis was compared with the meiosis of the haploid *A. esculentus,* with E = 65. A meiotic metaphase plate of the F_1 is figured in Fig. 3 for comparison with the haploid *A. esculentus* in Fig. 2. In 100 P.M.C.s in the two haploids, the number of bivalents per cell ranged from 0 to 2, the average being 0.67. However, the hybrid during meiosis showed a range of 0 to 7 bivalents per cell, the average being 1.63. This dissimilarity in chromosome association in the haploid and the hybrid is suggestive of dissimilarity between the genomes Y and F, since the T genome is similar in them both. Such dissimilarity may, however, represent the consequences of stabilising selection in the polyploid (cf. Riley, 1965).

Forty-five viable seeds were obtained from the F_1 plants treated with colchicine, and from these, amphiploid plants were established. The salient morphological features of the reproductive parts of the amphiploid (*tuberculatus* × *ficulneus*) and the two corresponding parental species are given in Table 1.

The amphiploid resembled *esculentus* in epicalyx number, petal colour, petal-spot colour and percentage of apparently good pollen. The fruit length was less than 6 cm in the parents, the hybrids and the amphiploids; in this respect, therefore, the amphiploid resembled only the short-fruited forms of *esculentus*.

Meiosis in the amphiploid was regular (Fig. 4), an average of 64.9_{II} per cell being formed. Thus, both in respect of chromosome number and meiotic behaviour, the amphiploid was similar to *esculentus*. However, the amphiploid did not set viable seeds, either on selfing or crossing with *esculentus*. There is thus some disharmony between its constituent F and T genomes.

Hybrids of A. esculentus with A. ficulneus and A. moschatus

These hybrids, with effectively a triploid constitution (65 + 36 chromosomes), could be obtained only through ovule and embryo culture (Gadwal, Joshi and Iyer, 1968). They both turned out to be totally seed-sterile. Chromosome association at meiosis in these hybrids is given in Table 2 and is illustrated in Figs. 5 and 6.

It will be seen that the maximal frequency of bivalents per cell in both these hybrids fell short of 36. Pairing in the *ficulneus* hybrid was substantially greater than in the *moschatus* hybrid (an average of 27.5 bivalents as against 8.3). Thus it appears that the *A. ficulneus* genome F is more nearly related to the unknown Y genome of *A. esculentus* than is the M genome of *A. moschatus*. Nevertheless, on grounds of incomplete pairing, the F genome must be regarded as less closely homologous with the Y genome than is the T genome of *A. tuberculatus* with the T' of *A. esculentus*.

Variation in fruit morphology

The varieties of *esculentus* used for study included a large range of variation for

Table 1 *Salient morphological features of amphiploid* (tuberculatus × ficulneus) *and its parents*

Character	*tuberculatus* I.W. 130	*ficulneus* I.W. 131	F_1	Amphiploid
Epicalyx no.	10−12	5−6	7−9	9−12
Petal colour	Yellow	White	Pale Yellow	Yellow
Petal-spot colour	Scarlet	Pink	Purple	Purple
Pollen stainability (%)	100	100	41	100
Pollen size (μm)	114−143	114	−	143−171
Fruit surface	Tuberculate	Smooth	Tuberculate	Tuberculate
Seed size	Small	Small	*	*
Seed colour	Black	Black	*	*

I.W. = Indigenous wild

* All seeds were small and black but none was viable.

Table 2 *Chromosome association at metaphase I in F_1 hybrids of* A. esculentus *(n = 65) with n = 36 chromosome species*

Valency per cell	I	II			III	P.M.C. no.
		Rods	Rings	Total		
1. F_1 (*esculentus* × *ficulneus*): 3x = 101 = Y + T′ + F						
Mean	45.7	22.1	5.4	27.5	0.1	10
Range	42−49	20−25	3−7	26−28	0−1	
2. F_1 (*esculentus* × *moschatus*): 3x = 101 = Y + T′ + M						
Mean	84.4	−	−	8.3	−	10
Range	69−95	−	−	3−16	−	

Table 3 *Range of fruit length and seed weight in* A. esculentus *and related taxa*

Character	*A. esculentus*	*A. tuberculatus*	*A. moschatus*	*A. ficulneus*
Fruit length (cm)	4.0−29.0	3.0−5.5	4.5−6.6	1.9−3.2
Seed weight (mg)	33−59	22−26	18	17

different characters occurring in the collections built up at the Division of Plant Introduction of Indian Agricultural Research Institute. The chief difference between *A. esculentus* and the wild Indian species is in the size and character of the fruits. *A. esculentus* has been selected for large, palatable fruits, and there is consequently both a much greater range and a much higher upper limit to fruit size in the cultivated species than in its wild relatives. Seed size also is much greater, probably in part as a side effect of selection for fruit size, but probably also because large seeds have a selective advantage under cultivation. The range of fruit length and seed weight in the species studied is set out in Table 3.

The fruit surface in a majority of the cultivated types of *esculentus* was plane. However, in one collection from the U.A.R., and one from Punjab, the surface was mildly tubercular. In EC 25622 from Sudan, it was almost as tubercular as in *tuberculatus*. In another culture from Sudan, EC 26886 received under the name 'Toti', the fruit was plane but it had a number of sparsely dispersed, non-raised, rhomboid markings. It seems likely, therefore, that the tuberculate character contributed to the polyploid by *A. tuberculatus*, has been lost in the course of improvement under cultivation, except in a few primitive cultivars.

Discussion

From the data presented in this paper we may conclude that the chromosome number of the commonly cultivated *A. esculentus* is n = 65, and that of this number a subgenome of 29 chromosomes is homologous with the 29-chromosome complement of *A. tuberculatus*. The homology of the remaining 36-chromosome genome is not yet clear. Of the two 36-chromosome species investigated, the relationships of *A. moschatus* with *A. esculentus* appear to be remote. Pairing was very limited in the *A. esculentus* × *A. moschatus* cross, and attempts to make the *A. tuberculatus* × *A. moschatus* cross have failed. *A. ficulneus* gives evidence of somewhat closer relationships. Considerable chromosome pairing occurs in the *A. esculentus* × *A. ficulneus* F_1. The cross *A. tuberculatus* × *A. ficulneus* is readily made. Some pairing occurs in the F_1, but the synthetic polyploid derived from it gives about normal pairing. However, the synthetic polyploid is not genetically balanced, as it has proved to be completely sterile. There remains, therefore, the problem of finding a 36-chromosome species more closely related than *A. ficulneus* to the 36-chromosome subgenome of *A. esculentus*.

Three lines of further study may be suggested. The first is to search for the *esculentus* types studied by Teshima (1933) and Ford (1938). If indeed there are cultivars with n = 36 or n = c. 33, they would appear to be the most likely relatives of the n = 65 forms. Such a situation is not unknown, but if it were demonstrated it would then be necessary to determine whether the two chromosome races should be regarded as distinct species as in the cottons (Hutchinson, *et al.* 1947), or as intraspecies chromosome races as in *Tradescantia* (Anderson and Sax, 1936).

The second line of study is to investigate the African species of *Abelmoschus*. The views of Vavilov (1951) and Murdock (1959) have been quoted. The two cultures received from Sudan (EC 25622 and Toti) showed, as stated earlier in this paper, some interesting primitive characteristics in the morphology of the fruit. It is also not unlikely that *tuberculatus*, as well as other wild species of *Abelmoschus*, may be found in the Sudanese, Nigerian and adjoining regions in Africa.

The third line is to carry further the study of the *Abelmoschus* species of Asia. On this a beginning has been made with the collection of the Indian species *A. crinitus, A. tetraphyllus* and *A. pungens*. Their cytological examination revealed the existence of yet another new chromosome number in *Abelmoschus*: n = 69. Studies with some of these species will be reported elsewhere. *A. angulosus* is distributed at high altitudes (above 1500 m) in the Nilgiri Hills of southern India.

Grown in Delhi, it does not come into flower. Chromosome number for this species collected from Sri Lanka was reported by Ford (1938) to be c. $2n = 56$. This will need confirmation.

A stage has thus been reached where one has to look elsewhere (i.e. outside India) for 36-chromosome species of *Abelmoschus*, some of which may be descendents of those involved in the evolutionary origin of cultivated *esculentus*. Another important region reporting a number of wild species of *Abelmoschus* is the area of Thailand, the Vietnams, Laos, Cambodia, the Philippines and west Irian (Waalkes, 1966). A search for more *Abelmoschus* material there would also be worthwhile.

Summary

Two haploids of *Abelmoschus esculentus* showed 65 chromosomes in their pollen mother cells, confirming $n = 65$ as the chromosome number of this species. Additional evidence that this set of 65 chromosomes includes 29 homologous with those of *A. tuberculatus* ($n = 29$), designated T, came from the nature of chromosome pairing in the F_1 of *A. esculentus* $\times$ *A. tuberculatus*. The remaining set of 36 chromosomes (designated Y) of *A. esculentus* exhibited very little homology with *A. moschatus* in the *A. esculentus* $\times$ *A. moschatus* cross. The cross of *A. esculentus* $\times$ *A. ficulneus* showed greater, but still very incomplete, homology. Supporting evidence for this inference came from: (i) comparative chromosome pairing during meiosis of haploid *A. esculentus* ($T' + Y$) and F_1 (*A. ficulneus* $\times$ *A. tuberculatus*) ($T + F$), and (ii) sterility of the colchicine-induced amphiploid (*A. tuberculatus* $\times$ *A. ficulneus*) on selfing or on crossing with *A. esculentus*.

Thus one of the parents of *A. esculentus* ($n = 65$) may have been like *A. tuberculatus* ($n = 29$). However, the other, which presumably had $n = 36$, was not like either of the two Indian species with $n = 36$, *A. ficulneus* and *A. moschatus*. In order to obtain a fuller picture of evolutionary origin of *A. esculentus*, the need to explore for 36-chromosome *A. esculentus*-like taxa in particular, because they were reported in earlier literature, and to determine fully the present-day geographical distribution of taxa which might have contributed the 36-chromosome genome to *esculentus*, was pointed out.

Acknowledgements

The authors are grateful to Dr H.B. Singh, Head, Division of Plant Introduction and Shri S. Ramanujam, Co-ordinator, All-India Project on Pulse Breeding at Division of Genetics of I.A.R.I.; and to Dr K.L. Mchra, Head, Division of Plant Improvement at the Central Grassland and Fodder Research Institute, Jhansi, for going through the manuscript and making valuable suggestions.

Solanum nigrum L.

S.L. TANDON
Department of Botany, University of Delhi and
G.R. RAO
Department of Botany, Aligarh Muslim University

The genus *Solanum* consists of approximately 1500 species (Bailey, 1949), mostly herbs or shrubs, with a very wide distribution ranging from the tropical to the temperate regions of the world and extending from sea level to an altitude of 2500 m. The species are broadly classified into two major groups, tuberiferous and non-tuberiferous. The tuber-forming species have attracted the attention of cyto-geneticists and plant breeders from time to time (Magoon, Ramanujam and Cooper, 1962) because of their great economic importance. Comparatively little attention has been paid to the cytogenetics and improvement of the less well known but still economically important non-tuberiferous forms of this genus.

The species of the *Solanum nigrum* complex are used in oriental medicine. The fruits and juices are used to cure stomach ailments, fevers, and blood impurities. The young shoots are also used in curing skin diseases (Chopra, Nayar and Chopra, 1956). In pharmacology the correct identification of the species and races used and an understanding of their relationships, are of great importance. Hitherto, identification has depended on morphological criteria, and the cytogenetic relationships of the forms used have not been elucidated. This is a prerequisite to the exploitation of the vast genetic variability available for the improvement of the quality and quantity of their drug content. The present paper deals with the cytogenetic relationships of the taxa included in the *Solanum nigrum* and *Solanum luteum* complex.

The material studied came from naturally growing Indian populations of *Solanum nigrum,* and a stock of *S. luteum* raised from seed supplied by Professor C.B. Heiser of Indiana University, U.S.A. Meiosis was studied in squashes of pollen mother cells fixed in Carnoy's fluid and made permanent with butyl alcohol (see Swaminathan, Magoon and Mehra, 1954, and Bhaduri and Ghosh, 1954*b*).

The naturally occurring forms

The Indian material was classified, mainly on the basis of fruit colour, into three morphologically distinguishable forms, and on cytological examination these were shown to be members of a polyploid series. Type I, with shining bluish-black fruits, is diploid (n = 12). Type II, with orange-red fruits, is tetraploid (n = 24). Type III with purplish-black, larger fruits, is hexaploid (n = 36). All these behaved normally in meiosis, giving bivalents only. Typical metaphase plates from the Indian types are illustrated in Plate 1*a, b, c.* Data on chromosome association at metaphase are given in Table 1. The metaphase configuration in *S. luteum* was indistinguishable from that of Type II (Indian tetraploid).

Table 1 *Type of chromosomal association at metaphase I in S. nigrum complex*

Type	Cytotype	P.M.C.s studied	Bivalents per cell		Univalents per cell	Trivalents per cell	Quadrivalents per cell	Chiasmata	
			Rings	Rods				per cell	per bivalent
I	Diploid	100	9.31	2.50	0.28	–	–	21.17	1.76
I × II	Triploid	50	4.40	1.60	21.46	1.00	–	13.10	0.72
II	Tetraploid	50	23.08	0.92	–	–	–	47.52	1.98
II × III	Pentaploid	20	18.50	3.50	15.01	0.33	–	36.20	1.20
III	Hexaploid	50	34.56	1.44	–	–	–	60.56	1.68
2(I × II)	Synthesised hexaploid (C1)	50	28.80	4.00	4.50	0.48	0.10	60.03	1.75
2(I × II)	Synthesised hexaploid (C2)	50	30.55	3.20	3.20	0.32	0.08	63.60	1.76
III × 2(I × II)*		50	33.90	1.52	1.16	–	–	68.32	1.92

* Naturally occurring hexaploid × synthesised hexaploid (CI)

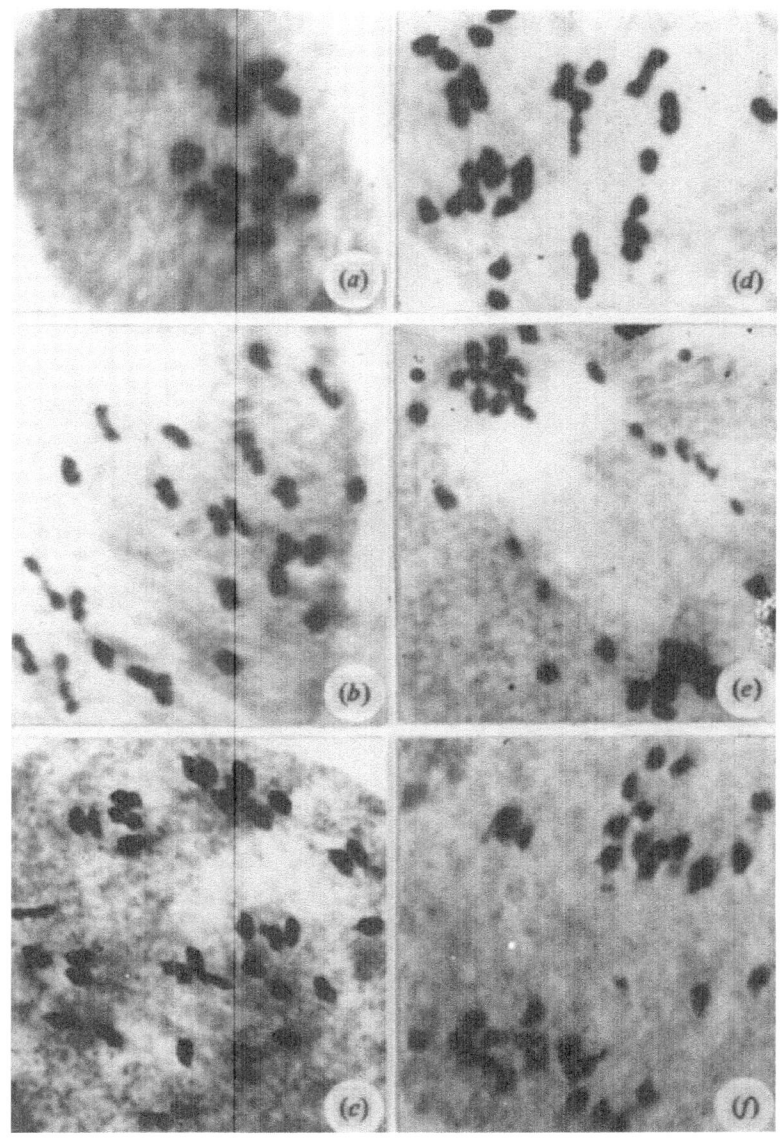

PLATE 1

(a–c) Metaphase I in diploid, tetraploid and hexaploid with 12 (× 1666), 24 (× 1666) and 36 (× 1500) bivalents, respectively.

(d–f) Meiosis in the triploid.

(d)   Metaphase I with 3 trivalents, 9 bivalents and 9 univalents (× 2133).

(e)   Anaphase I with numerous laggards (× 3028).

(f)   Anaphase I with 2 laggards and 17 chromosomes at each pole (× 2222).

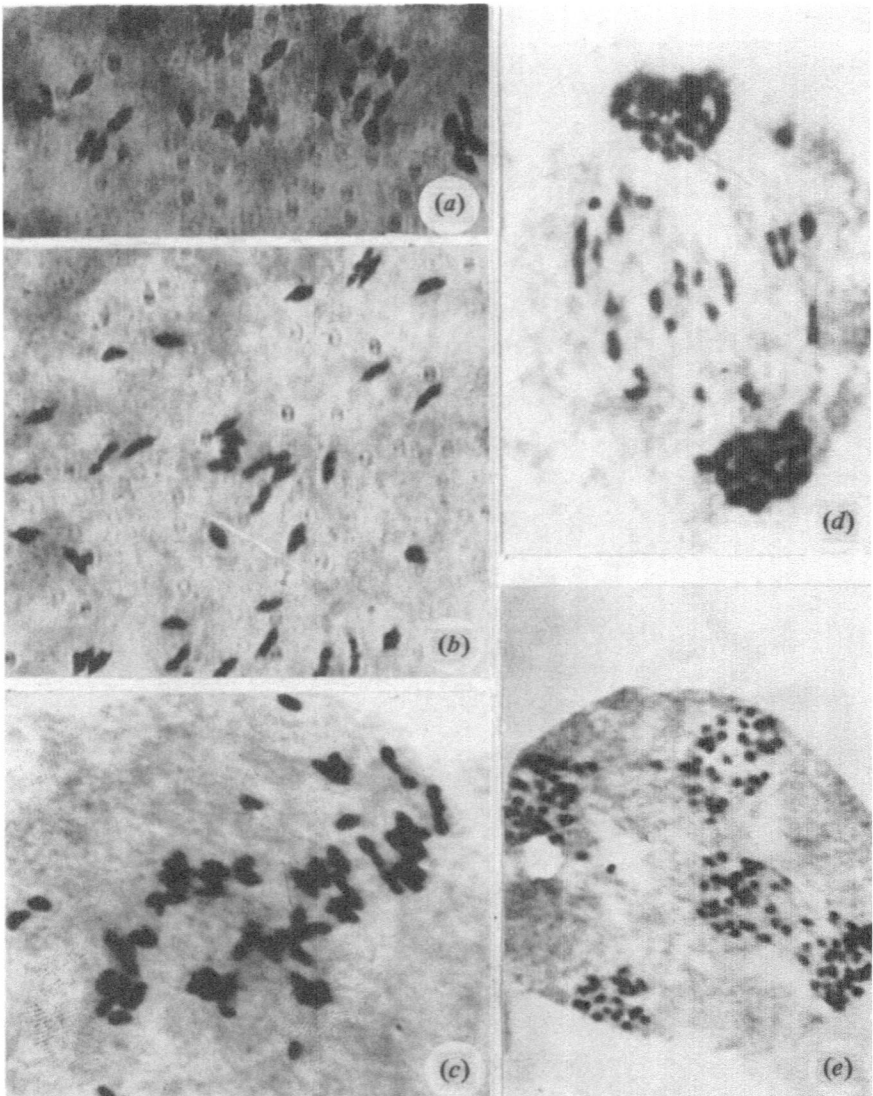

PLATE 2

(a)   Metaphase I in synthesised hexaploid (C1) with 36 bivalents (× 1481).
(b)   Naturally occurring hexaploid × synthesised hexaploid (C1). Prometaphase with 36 bivalents (× 1333).
(c–e) Meiosis in the pentaploid.
(c)   Metaphase I showing stickiness of the chromosomes (× 1444).
(d)   Anaphase I with laggards (× 1555).
(e)   Telophase II with 5 groups of chromosomes (× 1058).

Synthetic polyploids and hybrids

Tetraploids were synthesised by doubling the diploid with colchicine. Triploids were obtained by crossing diploid × natural tetraploid. Chromosome doubling was induced with colchicine, giving a synthetic hexaploid. Pentaploids were obtained by crossing tetraploid × hexaploid.

Crossing between races with different chromosome numbers was difficult. It was successful only when the parent with the higher chromosome number was the female, and even then fruit and seed set was a very small percentage of the large number of pollinations made. Crosses were also made between *S. nigrum* Type II and *S. luteum*, using the former as female parent. *S. luteum* resembles Type II morphologically, and is also a tetraploid. Crosses were successful, and in the hybrids pollen fertility was high.

The triploid hybrids of *S. nigrum* showed a wide range of meiotic irregularities. At diakinesis, in a majority of the pollen mother cells, large numbers of univalents were observed together with loosely paired bivalents (Plate 1*d*). At metaphase 1 a number of cells were observed to have 36 univalents (Table 1). However, in a few cells, occasional trivalents were also encountered. The equatorial region was occupied by a large number of univalents. In some of the cells, the univalents were present near the edge of the metaphase plate. In very rare cases univalents were observed to be at the poles. The univalents showed erratic behaviour at anaphase I, resulting in a large number of cells with irregular distribution of chromosomes at the poles (Plate 1*e*, *f*). However, a few cells were observed at anaphase I with an equal distribution of chromosomes. In a majority of cells lagging univalents were observed with varying frequencies (Plate 1*e*, *f*).

The hexaploid (C1) synthesised by doubling the triploid, was characterised by normal pairing of chromosomes. A number of pollen mother cells showed 36 bivalents at diakinesis (Plate 2*a*). However, a few cells showed occasional univalents, trivalents and quadrivalents. Data are given in Table 1. The synthesised hexaploid (C1) was crossed with the naturally occurring hexaploid. The progeny was at the hexaploid level with regular meiosis showing 36 bivalents in a great majority of pollen mother cells at both diakinesis and metaphase I (Plate 2*b*). Most of the bivalents were of the ring type. Some of the P.M.C.s showed univalents. Their number ranged from one to four. The frequency of occurrence of cells with univalents was, however, very low. Multivalents were absent. Data are given in Table 1.

The pentaploid obtained by crossing the naturally occurring tetraploid and hexaploid showed numerous meiotic irregularities. In a great majority of cases the chromosomes showed stickiness (Plate 2*c*). However, a few good stages of diakinesis and metaphase I were obtained for detailed analysis of chromosome pairing. At metaphase I the equatorial region was occupied by varying numbers of bivalents and univalents. The rest of the chromosomes were scattered all over the spindle. In very rare instances univalents were observed at the poles. Data are given in Table 1.

At anaphase I some lagging univalents were either in process of division or had already divided (Plate 2*d*). At telophase II, 60 per cent of the cells showed micronuclei in varying numbers. In some cells five groups of chromosomes were seen at early telophase II (Plate 2*e*). In very few cases normal tetrads were observed. Most of the sporads contained 5 or more cells.

Discussion

There have been conflicting reports on the nature of ploidy and meiotic behaviour of chromosomes in hexaploid *S. nigrum* (Magoon *et al.* 1962). Meiosis has been reported by some workers to be regular with 36 bivalents at metaphase I (Jørgensen, 1928; Bhaduri, 1933; Ellison, 1936; Swaminathan, 1949; Tandon and Rao 1966 *a, b*). These workers concluded that the hexaploid *S. nigrum* is an allopolyploid. Nakamura (1937) and Stebbins and Paddock (1949) reported multivalent associations during meiosis. Nakamura (1937) believed it to be an autopolyploid whereas Stebbins and Paddock considered it partly an allohexaploid. Stebbins (1950) and Günther (1959) considered this an autoallopolyploid. The present studies showed regular meiosis with normal pairing of chromosomes. This would indicate that hexaploid *S. nigrum* is an allopolyploid.

Chromosome pairing in hybrids provides an index of the degree of homology between the genomes thereby brought together. In the triploid hybrids only limited pairing occurred (average of six bivalents at metaphase), and pairing was only loose. Indeed, many cells with no bivalents (36 univalents) were seen and cells with little or no pairing were also observed by Chennaveeraiah and Patil (1968). Since no quadivalents were observed in meiosis in the tetraploid, autosyndesis is unlikely. Thus it is clear that the degree of homology between the diploid and the tetraploid genomes was low. These cytological observations led us to conclude that the diploid and the tetraploid are non-homologous. Their genomic constitution may be termed AA and BBCC respectively. The triploid may be represented as ABC. The triploid *S. nigrum* is completely sterile, and this substantiates the lack of genomic relationship between its diploid and tetraploid parents. This lack of relationship was confirmed by inducing autotetraploidy in the diploid and comparing with the natural tetraploid. The induced autotetraploid was an enlarged replica of the diploid and resembled it in all characters, including the colour of the berry. It in no way approached the natural tetraploid.

If the sterility of the triploid was chromosomal, doubling the chromosome number should restore fertility. Doubling was induced by colchicine treatment, and a study of the flower buds on induced hexaploid branches revealed full restoration of pollen fertility and fruit set. The synthesised hexaploid (C1: see Table 1) resembled the naturally occurring hexaploid in general pattern of morphological and cytological characters. But the raw nature of synthesised hexaploid is indicated cytologically by the occurrence of a few multivalents and univalents. The second generation hexaploid (C2: see Table 1) showed rather more regular pairing than the first generation.

The synthesised hexaploid crossed readily with the naturally occurring hexaploid producing fertile hexaploid offspring. Cytologically a few univalents and laggards at anaphase I were observed in some P.M.C.s of the offspring (Table 1), thereby indicating the existence of structural differences between their chromosomes.

These cytological studies indicate that the synthesised hexaploid is closely homologous with the naturally occurring hexaploid. Therefore, the genomic constitution of natural hexaploid may be designated AABBCC, as is also that of the synthesised hexaploid. The identification of the synthesised hexaploid with the natural hexaploid has been confirmed on the basis of chemical analysis of fruit pigment. The same anthocyanin pigment is present in the fruit skin of the naturally occurring hexaploid, and the synthesised hexaploid *S. nigrum.* The compound has been identified as petunidin-3-(*p*-coumoryl)rhamnosyl glucoside. It is also the fruit pigment of the diploid, thereby adding to the evidence for the view that the diploid is closely related to one of the parents of the naturally occurring hexaploid *Solanum nigrum.*

In the pentaploid hybrid which was obtained by crossing the naturally occurring hexaploid with the tetraploid, as many as twenty-two bivalents were observed in meiosis. Most of the other chromosomes were present as univalents. Thus it appears that out of five genomes of the pentaploid the chromosomes of four are paired and those of the fifth are in the form of univalents. Autosyndesis does not seem to occur, since quadrivalents are not observed in the tetraploid and hexaploid parents. The occurrence of as many as twenty-two bivalents in the pentaploid thus clearly indicates that two genomes of the naturally occurring hexaploid are genetically identical with the genomes of the tetraploid. Thus the tetraploid is closely related to the tetraploid parent of the naturally occurring hexaploid.

The Indian tetraploid *S. nigrum* and *S. luteum* resemble each other with respect to the very important diagnostic character, the colour of the berry. The chemical analysis of fruit colour showed the nature of the pigment to be carotenoid in both the forms. The identity of the naturally occurring tetraploid with *S. luteum* has been further confirmed by the production of fertile hybrid progeny with $n = 24$ chromosomes. The frequency of multivalents was low and most of the chromosomes formed bivalents at diakinesis and metaphase I, indicating the homology of the genomes of the two species. This indicates that naturally occurring tetraploid *S. nigrum* and *S. luteum* are genetically closely related.

Taxonomic implications of the genetical work

One of the important features of evolution in the *S. nigrum* complex is its partition into races which differ in ploidy, and between which crossing is consequently inhibited. In the present survey no natural hybrid was found, although the different cytotypes were seen to grow together. Within the races stability in genetic make-up has been assured through autogamy, and with high seed production and high viability the *S. nigrum* complex has the attributes for successful exploitation of weedy situations. The morphological differences and reproductive isolation between the races make it necessary to reconsider their taxonomic status. It seems clear that the three differing in ploidy should be accepted as distinct species. From the descriptions available it can be seen that, within each race, types from different parts of the world are morphologically similar. There is no reason, therefore, to make more than three species unless, within a chromosome race, cytogenetic differences were demonstrated.

Two names have been proposed for diploid races, *S. photeinocarpum* (Nakamura, 1937) and *S. nodiflorum.* Stebbins and Paddock (1949) considered that the two were conspecific, and that the proper name is *S. nodiflorum.* Unless cytogenetic differences are demonstrated, this name may be accepted as also including the Indian diploid race.

The identity of the Indian tetraploid race with *S. luteum* has been demonstrated morphologically and cytogenetically, and in the biochemical nature of the fruit colour. The name *S. villosum* has also been used for a tetraploid form, and the identity of this with *S. luteum* has not yet been demonstrated.

It has been shown above that the hexaploid is closely related to a hexaploid synthesised from a hybrid between an Indian diploid and an Indian hexaploid, in the terminology here suggested (*S. nodiflorum* × *S. luteum*). Westergaard (1948) deduced that the hexaploid may have arisen through amphiploidy of a hybrid of *S. nodiflorum* × *S. villosum*. If *S. villosum* may be equated with the tetraploids included under *S. luteum*, Westergaard's interpretation is the same as that proposed above. For the hexaploid the binomial *S. nigrum* may be retained.

Summary

Morphological and cytogenetical studies have been carried out on *Solanum nigrum* and *S. luteum* with a view to understanding their evolutionary history. Naturally growing populations of *S. nigrum* were classified into three distinguishable forms mainly on the basis of fruit colour and chromosome number.

	Race I	Race II	Race III
Fruit colour	Shiny bluish-black	Orange-red	Purplish-black
Chromosome number	$n = 12$	$n = 24$	$n = 36$

Meiosis was normal in all the three types, indicating the probable allopolyploid origin of the higher chromosomal forms.

Tetraploids were artificially produced from diploids by colchicine treatment and a comparison was made with naturally occurring tetraploids to see whether the latter originated by spontaneous chromosome doubling of the diploid. The induced tetraploid differed from the naturally occurring tetraploid in morphological as well as cytological characters. The dissimilar nature of the two was confirmed by an investigation of the chemical nature of the fruit pigment. Both cytologically and morphologically the naturally occurring tetraploid *S. nigrum* showed striking resemblances with *S. luteum*. The relationship between them has been confirmed by cytogenetical studies and an investigation of the chemical nature of the fruit pigment.

Triploids were synthesised by crossing the naturally occurring tetraploid with the diploid, The occurrence of a variety of meiotic abnormalities in the triploid showed that the three genomes in the triploid are dissimilar with respect to a majority of their chromosomes. The triploid was raised to the hexaploid condition by colchicine treatment. The synthesised and naturally occurring hexaploids are similar in morphological and cytological characters. Both are highly fertile and readily crossable. Their identity has been confirmed by a comparison of the chemistry of their fruit pigments.

The role played by the naturally occurring tetraploid in the origin of the naturally occurring hexaploid has been confirmed by the occurrence of as many as twenty-two bivalents in the pentaploid which was obtained by crossing the hexaploid with the tetraploid.

The foregoing cytogenetical studies of the *S. nigrum* complex lead us to believe that if the genomic constitution of the diploid is denoted as AA then that of the tetraploid is BBCC. The triploid will have a constitution of ABC. The genotype of the hexaploid will be AABBCC.

The polyploid races are characterised by well defined morphological characters and reproductive isolation from each other. In the light of these features it is proposed to give them specific rank.

Acknowledgment

We are grateful to Dr M.S. Swaminathan, Director, I.A.R.I., New Delhi, for his valuable suggestions received throughout the course of investigation.

4
Crops of the New World

Maize

K.R. SARKAR, B.K. MUKHERJEE, D. GUPTA and H.K. JAIN
Division of Genetics, I.A.R.I., New Delhi

Maize is one of the important food crops of India, occupying about 5.8 million hectares, largely in the northern and central states of the country. Production statistics for the last fifteen years illustrate the impact of scientific endeavour on the improvement of this crop. In 1955–6, total production for the country was only 2.60 million tonnes with an average yield of 704 kg per hectare; the corresponding figures for 1970–1 were 7.53 million tonnes and 1270 kg per hectare.

Local varieties collected from all parts of India show very limited genetic diversity, suggesting rather recent origin from a few introductions and offering little help to the breeder. Extensive introductions of exotic germ plasm have improved the situation and a number of high-yielding hybrid and composite varieties are now available.

Early history

The antiquity of maize in India is not clearly established. It is generally believed that the Portuguese introduced it to India from Europe during the early part of the sixteenth century. Little or no real evidence for the existence of maize on the Indian plains in pre-Columbian times is available from the literature, or from palaeontological or ethnological studies. No clearly established reference to maize is found in the Indian scriptures and epics, nor is the plant known to be associated with any religious or domestic rituals. There is not even an authentic Sanskrit name for the plant. The Sanskrit name, *Yaba-nala* (*Yaba* = barley, *nala* = reed-like), sometimes attributed to maize is also used for sorghum (Watt, 1892).

Watt (1892) discussed the common vernacular names for maize in various Indian languages and concluded that they do not throw any light on the history of maize in India. The most commonly occurring name, *Makkai* or *Makka,* which could mean 'from Mecca', suggests introduction from outside India. The other common name for maize in Indian languages is *Bhutta* or *Bhuta*. The origin of this word is obscure.

The earliest unambiguous reference to maize in India is found in the *Ain-i-Akbari,* the administration report of Emperor Akbar for the year A.D. 1590 (Watt, 1892). The indications are that maize was not a widely cultivated crop at that time, as it was not included in the long list of grains or pulses grown in India given by Abul Fuzl in the *Ain-i-Akbari.* It was mentioned only in connection with the description of another plant, 'Kewrah' (Watt, 1892).

The idea that maize might have been present in Asia in pre-Columbian times gained some credence with the discovery of 'waxy' maize in western China and its description by Collins (1909). By reviewing the references to maize in Chinese literature, he concluded that the plant existed in China for a long time and it came to the country from the West. However, his arguments do not conclusively prove the presence of maize in China before the sixteenth century. Anderson (1943, 1945) and Stonor and Anderson (1949) stimulated much interest in this topic. Stonor, from his experiences with primitive tribals of Assam and Burma hills, presented ethnological, linguistic, distributional and morphological evidence on maize varieties found there, and Anderson independently studied in detail the materials sent to him in the U.S.A. They concluded from their studies that there had been at least two major movements of maize in Asia, of which one was pre-Columbian and the other early post-Columbian. Anderson observed, in the same paper, that the complex of characters in Assam varieties is unknown in Mexican and Central American material, but they have some resemblance to those of Bolivia and the eastern lowlands. He concluded that maize had either originated in Asia or at least was present there before 1492. Mangelsdorf and Oliver (1951), however, contended that the Assamese maize of Stonor and Anderson was not at all unique and had close counterparts in South America. Mangelsdorf and Reeves (1959) cited the views of Merrill (1954) to conclude emphatically that the presence of all American plants in Asia was due to the early establishment of trade routes from Brazil to Goa in 1500.

The issue of the antiquity of maize in India is reopened by the recent survey of primitive germ plasm by the Indian Agricultural Research Institute. Dhawan (1964) reported that a wide spectrum of variability in maize exists in the northeastern Himalayas. At least two of the Sikkim collections, namely Sikkim Primitive 1 and Sikkim Primitive 2 (SP 1 and SP 2), presented strikingly primitive features and he described SP 1 in detail. It seems highly improbable that these are relatively recent introductions. Thapa (1966) recently pointed out certain ethnological and linguistic evidences in support of the view that maize might have been in this region prior to the discovery of America by Europeans.

The two Himalayan primitives, SP 1 and SP 2, have characteristics that are markedly different both from American primitives and from advanced commercial types (Gupta and Jain, 1971a). They are extremely late in maturity, silking about a month later than the advanced types. They are also shorter in height, lower in yield, having ears which are smaller in length and diameter, with fewer kernel rows and lower kernel weight compared to the evolved types. The American primitives, on the other hand, do not differ so markedly from the advanced commercial types. The Himalayan primitives, especially the SP 1, have a very low pollen diameter (66μm) compared with other maize varieties (83–85μm).

These characters of the Himalayan primitives, together with the emergence of ears from the upper joints of the stalks, reduced internode length, and the occurrence of male and female flowers in the same inflorescence, show that these varieties are closer to the progenitor corn plant reconstructed by Mangelsdorf (1958) than such American races as Chapalote, Nal-Tel and Palomero Toluqueno, which are believed to be its immediate descendants in the New World.

That some differentiation at the cytoplasmic level exists between the Himalayan primitive varieties and the advanced varieties of maize was brought out by observations on reciprocal hybrids between the Himalayan primitives and commercial varieties (Gupta and Jain, 1971b). Significant differences between reciprocal crosses for a number of characters including maturity index, plant height and yield components such as ear length, number of kernel rows and thousand-grain weight were observed with advanced cytoplasm favouring better expression of various characters excepting ear length. Similar cytoplasmic effects were also observed in maize by Bhat and Dhawan (1969).

Results of karyotype analysis of the SP 1 and SP 2 varieties (Jain and Gupta, 1971a) suggest that the Himalayan primitive varieties are quite distinct from present-day advanced types in their chromosome structure. The pachytene chromosomes in SP 1 and SP 2 are somewhat smaller in size than chromosomes of many of the American varieties of maize as reported by Longley (1939). It appears that the reduced size of the chromosomes in SP 1 and SP 2 is due to a greater degree of condensation, which, however, is not uniform in all the chromosomes. The observation that the arm ratio of the different chromosomes in these two primitive varieties and the evolved types is almost the same, supports this conclusion.

The knob-forming pattern in the two Himalayan primitives was found to be different from that of the evolved varieties. While the number of knobs on the chromosomes of the primitive varieties is not very different from that reported in many of the advanced types, the Himalayan primitives show a characteristic position of the knobs in the chromosomes. Thus, four of the knobs, 7L, 8S, 8L and 10La, all observed in SP 2, and 10LA observed in SP 1, appear to indicate new knob-forming sites not commonly known in evolved varieties of maize. The knobs in the Himalayan primitive varieties were found to be larger in size compared with those reported for most of the advanced varieties.

In another study (Jain and Gupta, 1971b), chiasma frequencies in six primitive varieties including SP 1 and SP 2 and three advanced types were determined. The primitive varieties, as a group, showed a significantly higher chiasma frequency than the advanced types, suggesting that the primitive varieties are quite distinct from the present-day varieties of commerce. The two would appear to have been separated from each other over a longer period than the concept of a wholly post-Columbian introduction of maize in Asia would suggest.

Comparison of the mitotic cycle of two primitive varieties, SP 2 (Himalayan) and Palomero Toluqueno (Mexican), and two advanced varieties, KT 41 (Indian) and Mexican June, presented some interesting findings (Jain and Gupta, 1971c). Observations on mitotic index and duration of cell cycle revealed that both the primitive varieties show a longer nuclear cycle compared with the improved varieties. The S phase, which represents the period during which DNA is synthesised in the chromosomes, is longer in the two primitive varieties. This suggests that the two primitive varieties have a greater DNA content in the cells compared with the two advanced varieties. It may be supposed that the cultivated varieties incurred some loss of DNA through evolution. The significance of this loss is open to conjecture.

These observations on the two Himalayan primitive varieties clearly establish them as distinct entities different from the advanced types as well as the American primitive types. Speculation on the origin and location of these varieties in remote Himalayan regions would be premature at this stage. Nevertheless, they open up an entirely new angle on the origin, evolution and distribution of maize.

It seems that the question of the presence of maize in Asia before 1492 remains open, but the protagonists have only one stick to wield. It must be admitted that the presence of primitive races in Sikkim, Nepal, Bhutan or Assam hills is extremely puzzling and cannot be explained on the assumption of introduction and spread of maize in the post-Columbian era. On the other hand, to quote Weatherwax (1954), there is not 'a single written record of undisputed pre-Columbian date which establishes the presence of the plant in Asia at that time... At present we are not aware of a single plant segment, artifact, illustration, written record, or other concrete evidence of the plant in Asia earlier than about seventy-five years after it could have been taken to Eastern India from Brazil...' It is unfortunate that insufficient attention has so far been given to a study of the germ plasm of the primitive varieties of the eastern Himalayas to estimate their antiquity. Nor have archaeological studies similar to those in Mexico been undertaken in this part of the world.

Most of the Indian maize material is of yellow and white flint types, and is named after the locality where the particular strain is prevalent. The limited genetic diversity of the Indian germ plasm has been shown by Bharadwaj (1960) and also by the work carried out in the Co-ordinated Maize Breeding Scheme of the I.C.A.R. Bharadwaj was not able to get substantial heterosis by crossing Indian X Indian types and confirmed Watt's (1892) conclusion that modern maize types in India may have originated from a few introductions.

Stonor and Anderson (1949) concluded that post-Columbian introductions of maize to Asian countries are essentially of Caribbean types. However, Mukherjee, Gupta, Singh and Singh, (1971) observed from metroglyph and index score analysis that the variability pattern of the Caribbean germ plasm was different from that of the Indian complexes. Disregarding the group containing the two Sikkim primitive varieties (SP 1 and SP 2), which is strikingly different, there are two morphologically distinct complexes from India proper. These, consisting of representative northern and peninsular varieties, resemble Mexican and Columbian germ plasm in variability pattern. More elaborate studies are required to confirm the source of introduction and evolution of the Indian cultivated varieties.

Maize improvement work in India

Scientific improvement of maize in India started only about twenty-five years ago. Prior to this, not much attention was given to this crop. Farmers cultivated maize and made their own visual selection for seed. Breeding projects to exploit hybrid vigour were taken up in the Punjab State Department of Agriculture in 1945 and subsequently at the Indian Agricultural Research Institute, New Delhi. Two hybrids, namely Punjab Hybrid no. 1 and a three-way cross hybrid were developed from indigenous open-pollinated varieties. These hybrids gave only 10—15 per cent more yield than the local maize varieties. Though some gain had been made, it was not so spectacular as that achieved by maize breeders in the U.S.A.

During the early phases of maize breeding, fifty-nine hybrids of different maturity groups were brought from the U.S.A. and Australia and were tested at several locations in the maize-growing tracts of the country. Several of the better adapted southern U.S. hybrids outyielded the local open-pollinated varieties by 80—120%. On the basis of their performance, the best yielding hybrids were produced and

124

distributed on a limited scale in Jammu and Kashmir, Punjab, Uttar Pradesh, Andhra Pradesh and Delhi. Although this was very encouraging, it was found that there was considerable difficulty in maintaining the parental inbreds of the U.S. hybrids in most parts of India. Moreover, the dent grain type of the U.S. hybrids was not acceptable to the Indian consumers.

With the organisation of the All-India Co-ordinated Maize Improvement Scheme in 1957, the work on improvement was considerably intensified, and has contributed significantly to all aspects of maize cultivation in India. In the initial stages a large collection of inbred lines, varieties and germ plasm complexes were introduced from the U.S.A., Mexico, Colombia and the Caribbean region. A large collection of local varieties was also made from different parts of the country. The greater part of the local collection came from the northern and northwestern states of India. A large-scale testing programme with this material was carried out at the thirteen regional stations of the Co-ordinated Scheme. Inbred lines numbering more than 4000 were developed from Indian and adapted exotic germ plasm. A large number of experimental single crosses were made and their double-cross performance was predicted. All the inbred lines were screened for disease and pest resistance. Detailed accounts of parental material, derived inbred lines, and hybrids and composites released commercially, are given in the Annual Reports of the All-India Co-ordinated Maize Improvement Scheme.

By 1961, four outstanding double-cross hybrids for the major maize-growing areas in the country were released by the Co-ordinated Project for cultivation. These were: Ganga 1 and Ganga 101 for the northern plains; Ranjit for south Rajasthan, Gujarat and Maharastra, and Deccan for peninsular India. In 1962, a very high-yielding hybrid VL 54 was developed by the Vivekanand Laboratory, Almora and released for the U.P. hills. In the following years, a series of better hybrids was developed, such as Ganga Safed 2 for white maize-growing areas, Hi-Starch for the starch industry, Ganga 3, an early maturing hybrid to replace Ganga 1 and Himalayan 123 for the hilly regions. In 1968, a new hybrid Ganga 5 possessing not only high yield potential but also marked resistance to the brown stripe downy mildew (*Sclerophthora rayssiae*) and the leaf blights, was issued, replacing Ganga 3 which was found to be highly susceptible to brown stripe downy mildew. Eleven hybrids have been released in the decade 1961–71. The contribution which they have made to India's agricultural economy is shown in Table 1.

Studies in Central and South America and India on intervarietal and inter-racial crosses have demonstrated that heterotic effects for yield and other characters of a high order can be obtained by crossing varieties of wide genetic diversity. Such varietal crosses have given yields comparable to, or higher than, the best double-cross hybrids developed so far. Advanced generations of these crosses and of populations developed from multiple crosses of elite varieties, which have been designated composite varieties, have been shown to give stable yields at a high level of productivity. Based on this concept, considerable work has been carried out in the Co-ordinated Maize Improvement Scheme and this has led to the development and release of six high-yielding composites, namely Amber for peninsular India and the Hamalayan hills, Jawahar for the northwest plains and peninsular India, Kisan and Vikram for the northern plains, and Sona and Vijay for the northwest plains. Most

Table 1 *Gains realised from high-yielding varieties of maize in the agricultural economy of India* *

Total area under maize (1970–1)	5.84 million hectares
Total production	7.41 million tonnes or Rs. 444.0 crores
Average yield under high-yielding varieties	3500 kg/hectare
National average	1270 kg/hectare
Gain	+ 2230 kg/hectare
Additional gain realised in a year due to the high-yielding varieties programme (10% of the cropped area under improved varieties)	0.6 × 2.23 = 1.338 million tonnes
Value of the gain @ Rs. 600/ton	Rs. 80.3 crores

* Courtesy of Dr Joginder Singh, Co-ordinator, All-India Co-ordinated Maize Improvement Scheme.

of these composites possess the same yield levels as the released double-cross hybrids, ranging from 5000 to 7000 kg per hectare. These composites have also been used as sources for the derivation of better inbred lines from which to develop higher-yielding double-cross hybrids.

Special objectives

Investigations on the development of sweet corn and pop corn varieties and the improvement of protein quality were also initiated in the course of the Co-ordinated Maize Improvement Project. It was found that accessions of exotic sweet corn germ plasm did not have sufficient vigour to be suitable for direct use under the conditions prevailing in this country. In an effort to build up vigour and simultaneously to maintain some of the qualities of the sweet corn, a crossing and selection programme was started. It has been possible to establish many of the qualities of sweet corn with flint and dent bases where the yield appeared to be satisfactory.

Work on the improvement of the quality of maize by using the opaque-2 gene was initiated in 1965 by N.L. Dhawan, Joginder Singh and their collaborators. A number of genetic stocks carrying opaque-2 were obtained from the U.S.A. These stocks were, however, very poor in vigour and could hardly be maintained. Using them as sources, the opaque-2 gene was incorporated in four promising adapted stocks, namely J_1, Cuba 11J, Doeto and Antigua Gr. 1. After the second backcross generation, the four recovered populations were intercrossed and the six possible combinations were composited. The advanced generations of the composite and the advanced generation of the recovered J_1 have been released under the names Shakti and Ratan respectively. These contain lysine as high as 3.45 g per 100 g of protein. Feeding experiments carried out comparing the normal hybrid Ganga 3, and casein from milk, with Shakti suggested that the protein efficiency ratio was 1.20, 2.08 and 3.38 for Ganga 3, casein and Shakti respectively (*Progress report Co-ordinated Maize Improvement Scheme*, 1972). This indicated the superior nutritive value of the variety Shakti. Another high-lysine composite, Protina was developed by the Uttar Pradesh Agricultural University and released concurrently with Shakti and Ratan under the auspices of the Co-ordinated Maize Improvement Project. Opaque-2 has also been incorporated into a number of other elite composites and promising inbred lines of released and experimental hybrids and these materials are being tested.

The current maize improvement programme in India is directed towards locating new sources of superior genes for enhancing yield and for combating diseases and pests, and mobilising these genes into elite populations. In order to achieve these objectives, new germ plasm from different parts of the world is being incorporated into the breeding programmes. At present, greater emphasis is being placed on plant type. There are good indications that the stalk-ear ratio in maize can be changed drastically. Some of the brachytic and other dwarfs produce a normal-sized ear on a plant which is hardly four feet tall. By utilising this material and making appropriate changes in plant population, it may be possible to raise the yield well beyond the present limit.

Pathologists and entomologists working in association with breeders have made a considerable contribution in screening and developing varieties resistant to diseases and pests. A number of inbreds have been found to be resistant to diseases such as leaf blight, brown stripe downy mildew, brown spot and the various stalk rots. Recurrent selection programmes for the improvement of released composites with respect to disease resistance have resulted in highly resistant reconstituted populations and these are currently under test (*Progress Report, Pathology Section, Co-ordinated Maize Improvement Scheme*, 1971).

Screening for stem borer has shown that Antigua Gr. 1 is the most resistant material available. A number of other stocks have also been found to possess some degree of resistance. More elaborate field trials are being carried out by the Co-ordinated Scheme to obtain stable sources of resistance (*Progress Report, Entomology Section, Co-ordinated Maize Improvement Scheme*, 1971).

Current varieties carrying the opaque-2 gene have poor storage characteristics. These are improved when the gene is transferred to strains with flint-type grain. Varieties with flint/opaque-2 grain are in course of development.

Advances in maize production so far have come from the exploitation of the germ plasm of the varieties established on the Indian plains, combined with the germ plasm of exotic varieties. In recent years the Genetics Division of the Indian Agricultural Research Institute has initiated a scheme to collect and study the very different germ plasm resources of the northeastern Himalayan region. A large number of collections has already been made and this material is being evaluated (personal communication from Dr Bhag Singh). It is expected that the detailed study of the material from the northeastern Himalayas will also contribute significantly to our understanding of the problem of antiquity of maize in India.

Acknowledgments
The authors are thankful to Dr D. Sharma of the J.N.K. Vishwavidyalaya and Mr J. Thapa of Sikkim for supplying information on the primitive maize in Sikkim.

Grain amaranths

MOHINDER PAL and T.N. KHOSHOO
National Botanic Gardens, Lucknow

Recent archaeological studies in South and Central America reveal that grain amaranths (genus *Amaranthus*) have been associated with Man since prehistoric times and are among the most ancient crops, having been cultivated as early as 4800 B.C. (Agogino, 1957). Although considerably overshadowed by cereals, they are still cultivated in several countries in the Americas, in Asia and in some pockets in Africa. India is a country where these grains are cultivated to a substantial extent. Generally in India they are used as a subsidiary food, but in some places in the western Himalayas at elevations above 1800 m they are grown as a regular crop in infertile and stony soil and constitute the staple food in place of wheat.

The seed of the grain amaranths is very small but it is comparable to the improved cereals in mineral, carbohydrate, fat and protein contents (Misra, Pal, Mitra and Khoshoo, 1971). The protein quality is better in as much as it is richer in certain essential amino acids including lysine and methionine, and the protein efficiency is comparable to casein (Subramaniam and Srinivasan, 1952). Furthermore, grain amaranths are very hardy, suitable for growing in areas with a short season and deficient soils, where cereals cannot be grown with ease. Thus they have potentialities as a subsidiary food and may play an important role in feeding the hungry world.

There are four species of grain amaranths, *A. hypochondriacus, A. cruentus, A. caudatus* and *A. edulis.* Associated with these are three weedy species, *A. hybridus, A. powellii,* and *A. quitensis.* Although most of the grain species are cultivated in several parts of the world, only one weedy species, *A. hybridus,* is now widely spread. Ecogeographical, ethnobotanical and archaeological data show that all the seven species belong to the Americas, where they have been firmly entrenched in the lives of prehistoric peoples who cultivated them since time immemorial. Among other plants, they represent the counterpart of the common cereals of the Old World. In Mexico amaranth has been an important grain crop since 5000–3000 B.C., and in the Aztec empire Moctezuma received an annual tribute of 200 000 bushels of amaranth seed, which was equal to the maize tribute. Even in the sixteenth century *A. hypochondriacus* was a major grain crop in the tropical highlands of central Mexico and was a staple diet in the region at the time of the conquest. The decline in production from post-Colombian times is due partly to the arrival of the Old World cereals and partly to suppression by the Spaniards because of the important role of these grains in Mexican religious ceremonies (see Purseglove, 1968; Sauer, 1950, 1967).

In India *A. hypochondriacus* is cultivated, and the seed used in a variety of ways.

PLATE 1.
Round cakes made with popped amaranth grains and jaggery as sold in
Lucknow markets.

In places where it is grown for local consumption and forms the staple diet of the
people (the Sutlej valley and higher elevations in Kumaon and Gharwal), the seeds
are used in a way similar to wheat, being first ground into flour and then made into
a dough with water. From the dough thin cakes or *'chapatis'* are prepared by patting
between the palms, and baking over a fire. The grains may sometimes be popped
before grinding (Singh, 1961). In Nepal (Singh, 1961) the parched seeds are ground
to flour and eaten as gruel (*'sattoo'*) with milk or water. The popped grains are put
in milk and the porridge thus made is taken in most parts of India on fast days, as
the cereals are not allowed. However, the most common use of these grains in India,
in which a major part of the produce is consumed, is in the form of sweetmeats. The
popped grains are mixed with jaggery and converted into balls or cakes. In Lucknow
(U.P.), for example, round cakes of ½ m diameter and 10 cm thickness are prepared
which are then cut up and sold in portions (Plate 1).

Characters of the grain amaranths
From a comparison of the domesticated and weedy species, an idea can be formed
of the important characters that led to the establishment of the former as grain
crops (see also Sauer, 1967). The monoecious habit, together with predominant
outbreeding, is an important feature of the grain amaranths which appears to have
helped in their domestication. In the grain amaranths each glomerule (the basic
unit of the inflorescence) contains one male flower and about 250 female flowers.
Evidently, the higher number of female flowers in Section *Amaranthus* has been an

130

PLATE 2.
Two-week-old seedlings of *A. caudatus* (left) *A. hybridus* (right) and
their F₁ hybrids (middle four). The hybrid seedlings do not proceed
beyond this stage.

advantage in its exploitation for grain.

The presence of very large, closely set, compound terminal inflorescences with
many branches held well above the leafy stems in the grain amaranths in comparison
to the small axillary or laxly branched, relatively very small terminal inflorescences
in Section *Blitopsis,* have made greater grain yield possible, although seed size has
not increased. It is usual to produce several tens of thousands of seed and yields of
500 000 seeds per plant have been reported (see Sauer, 1967) even in the wild species
like *A. retroflexus* of Section *Amaranthus.*

The relatively short and weak bracts in the grain amaranths have facilitated selection of inflorescences that are less stiff and prickly when rubbed between the
hands, which is the common method of extracting the seed. This is in contrast to
the coarse and often prickly inflorescences in their weedy counterparts and other
amaranths. The presence of a dehiscent utricle allows easy threshing and winnowing
in comparison to the indehiscent utricles in several species of the genus, including
some populations of ancestral species such as *A. powellii.*

The pale ivory seed of the grain types is preferred for its popping qualities and
flavour in comparison with the black seeds in weeds. Black seed is a dominant
character and black-seeded plants are constantly weeded out whenever crossing
between them takes place. Brightly coloured inflorescences — reds, yellow and
clear greens — have been preferred in the grain types over the usually green with

131

dull red pigmentation of the weedy species, because of the religious significance attached to them.

Theories of origin

Early botanists and explorers were struck by the excellent establishment and the widespread distribution of grain amaranths and some of their progenitors in Africa and Europe, and in particular in Asia, so much so that they thought them to be indigenous to Asia. De Candolle (1886) and Vavilov (1951) (see also Darlington, 1963) believed that grain amaranths like *A. cruentus* (*A. paniculatus*) arose in the Indo-Burma region. However, based on data from the ecogeographical, morphological, archaeological, ethnobotanical and philological studies of Sauer (1950, 1967), it has been concluded that the domestication of grain amaranths began in the Americas with the dawn of American Indian agriculture and there is evidence for the amaranths being one of the most ancient (4800 B.C.) American crops (Agogino and Feinhandler, 1957; Agogino and Hester, 1958; Purseglove, 1968). There is no valid evidence of the movement of crops by man between the Old and the New World in pre-Columbian times (Purseglove, 1968).

Most of the grain amaranths reached Asia in the eighteenth century and early in the nineteenth century *A. hypochondriacus* became widespread as an important grain crop in India where at present it is cultivated to a greater extent than elsewhere in the whole world. Associated with it is its probable weedy ancestor, *A. hybridus,* and occasional hybridisation between the two in the Sutlej valley has been observed. However, the purity of the crop is maintained through selection for white seeds. That this weed was perhaps introduced with the cultigen is apparent from the fact that *A. hybridus* has not been recorded in *Flora of British India* (Hooker, 1884). The other two wild relatives of the grain species, namely *A. powellii* and *A. quitensis,* have not been reported from India. Small patches of *A. caudatus* have often been seen in the fields where maize and *A. hypochondriacus* are grown as regular crops. However, in contrast to the latter, the authors have not noticed regular agricultural cultivation of *A. caudatus* anywhere in India. When found in agricultural fields, it may be an escape from its primary use as an ornamental in India. Very recently a third species, *A. edulis,* has been introduced in Himachal Pradesh (Singh and Dadlani, 1967) and Lucknow (present authors). It seems to have adapted itself well under both agroclimatic conditions.

The grain amaranths, although an accepted food on religious days of the Hindus, do not have a Sanskrit name, but such names exist for vegetable amaranths indigenous to southeast Asia and India (C.S.I.R.I., 1948; Aiyer, 1956; Parkash, 1961). Furthermore, the morphological and genetic evidence obtained by the authors shows that the four domesticates collected from the Old and the New World are perfectly conspecific and no differences, suggestive of any genetical and morphological divergence, exist. However, the same cannot be said of other non-cultivated weedy amaranths. In view of this we have to look to the New World for the origin of grain amaranths.

Essentially two hypotheses have been suggested by Sauer (1967). The first envisages the occurrence of three complexes, each containing one domesticate and one weedy species. The weedy ancestors were useful plants to primitive man in the

Table 1   *'Semifertile' natural crosses in Amaranthus postulated from observational data*

	A. hybridus	A. hypochondriacus	A. powellii	A. cruentus	A. quitensis	A. caudatus	A. edulis	
A. hybridus								C Covas (1950)
A. hypochondriacus								S Sauer (1950, 1967)
A. powellii	S,T	S						T Tucker and Sauer (1958)
A. cruentus	S	S,Si	S					Si Singh (1961)
A. quitensis	S			S?				
A. caudatus	T		T					
A. edulis	C							

three areas and each grain species arose independently of the others through selection by different prehistoric peoples. Thus arose *A. hypochondriacus* from *A. powellii* in North, *A. cruentus* from *A. hybridus* in Central and *A. caudatus* and *A. edulis* from *A. quitensis* in South America.

Sauer suggests alternatively that it is possible that each grain species arose by domestication from the same wild ancestor. Variation within each of the domesticates could be explained on the basis of introgressive hybridisation. For example, in *A. quitensis,* the semi-cultivated types with intense red-coloured inflorescences used for colouring maize dishes look like *A. cruentus,* which is also used for the same purpose in North America. Such types, according to Sauer (1967), may be introgressants of *A. cruentus* into *A. quitensis.*

Based on his observations of the extensive variation both within and between the species of grain amaranths and their weedy relatives, Sauer postulated free and frequent intercrossing between both cultivars and weed species, but no experimental studies of crossability were made. Natural crossing as suggested by Sauer (1950, 1967), Covas (1950), Tucker and Sauer (1958), and Singh (1961) is tabulated in Table 1.

Cytogenetic evidence

The present authors have set out to test the cross compatability of these species experimentally. In the present work the four cultivated species, *A. hypochondriacus, A. cruentus, A. caudatus* and *A. edulis,* were available, but only two, *A. hybridus* and *A. powellii* (*A. bouchoni* component) of the three wild species. These taxa are included in a systematic programme of hybridisation.

Grant (1959) has published chromosome numbers of the grain amaranths, and these have been confirmed in the present study. In *A. hypochondriacus, A. caudatus, A. edulis* and *A. hybrides* n = 16; in *A. cruentus* and *A powellii* n = 17. There are no polyploids in the group. All species exhibit normal meiosis and pollen and seed fertility. The results of present hybridisation experiments are set out in Table 2.

Table 2   *Results of experimental crosses between* Amaranthus *species*

	A. hybridus	*A. hypochondriacus*	*A. powellii*	*A. cruentus*	*A. quitensis*	*A. caudatus*	*A. edulis*	
A. hybridus								
A. hypochondriacus	F							x failed
A. powellii	st	st						d seedlings died
A. cruentus		x						ab seedlings abnormal
A. quitensis	Not available							st F₁ normal but sterile
A. caudatus	d	ab						F F₁ partially fertile
A. edulis	d	ab		x		F		

In Table 2 the species are set out in an order that takes account of chromosome number as well as of geographical distribution. The two cultivated species *A. caudatus* and *A. edulis,* and the weedy *A. quitensis,* form a group with n = 16 chromosomes indigenous in the South American Andes. The Central and North American species can be divided on chromosome number. With n = 16 are the cultivated *A. hypochondriacus* and the wild *A. hybridus,* and with n = 17 are the cultivated *A. cruentus* and the wild *A. powellii.* This grouping differs from Sauer's in relating *A. hypochondriacus* to *A. hybridus,* and not to *A. powellii.* The crossing results so far available fit this grouping.

Two of the three possible within-group crosses have been made (*A. quitensis* was not available), and both gave normal F₁s which were partially fertile and produced F₂ progenies. Pachytene pairing showed a small amount of structural hybridity for small segments. Meiosis was normal with chiasma frequency lower than in the parents, leading to 25–55 per cent fertile pollen and 49–68 per cent threshable seeds (Pal and Khoshoo, 1972). The F₁ of *A. edulis* × *A. caudatus* resembled *A. caudatus.* The hybrid *A. hybridus* × *A. hypochondriacus,* on the other hand, was more or less intermediate in most characters. In the F₂s about 10–20 per cent of unthrifty seedlings appeared, and these died very early. The rest of the F₂ plants were healthy and vigorous. Plants quite near one or other parental phenotype were recovered as also were those that showed different degrees of recombination. In general, greater independent assortment of distinguishing characters was found in hybrids involving *A. edulis* and *A. caudatus* than in the other combination. Amphidiploids of *A. edulis* × *A. caudatus* and *A. hybridus* × *A. hypochondriacus* showed typical autoploid or segmental alloploid characters during meiosis (Khoshoo and Pal, 1972). It is evident that the parents of these hybrids are not strongly differentiated and gene exchange is possible between them.

The between-group crosses can be divided into those involving n = 16 chromosome species only, and those involving crosses between n = 16 and n = 17 chromosome species. Of the former, all four of the possible crosses (omitting *A. quitensis,* which

134

was not available) were made and produced hybrid seed. In the two crosses of
A. hybridus (*A. caudatus* × *A. hybridus* and *A. hybridus* × *A. edulis*) germination
of the hybrid seed was normal, but the seedlings died at about the time of emergence
of the first true leaf (see Plate 2). In the two crosses involving *A. hypochondriacus,*
the hybrid seed of *A. edulis* × *A. hypochondriacus* has a ruptured seed coat and
exposed endosperm. This not only affects the extent of germination but also subsequent seedling growth. The mature plants of this hybrid and to some extent those
of *A. caudatus* × *A. hypochondriacus* were stunted, showing splitting of stems
thereby exposing cortical region, peculiar twining in stems and inflorescence axes,
stem and root tumours, distorted leaves apparently showing a pathological virus-
like syndrome, and underdeveloped anthers which may not dehisce (Pal and
Khoshoo, 1972). During meiosis there was unmistakable evidence at pachytene of
structural hybridity with deletions, long or short differentiated segments and
inversions. Although bivalents were formed, they possessed chiasma frequency lower
than either of the parents. There was total pollen and seed sterility. In both cases
the hybrids closely resembled *A. hypochondriacus.* Of the latter, four crosses were
attempted between n = 16 and n = 17 chromosome species. The hybrids *A. powellii*
(*A. bouchoni*) × *A. hybridus* and *A. powellii* (*A. bouchoni*) × *A. hypochondriacus*
were successfully made and gave normal, but totally sterile, F_1 plants. The crosses
A. cruentus × *A. edulis* and *A. hypochondriacus* × *A. cruentus* failed completely.

Evidently the differentiation between the three groups of species is very great.
The data show that, contrary to general belief that, because of wind pollination,
hybridisation in amaranths is very common, a variety of isolating mechanisms (Pal
and Khoshoo, 1972) in fact prevent hybridisation. All these facts indicate that the
theories of introgression in the history of the grain amaranths put forward by Sauer
are not substantiated by the results of the experimental hybridisation done in this
laboratory.

Comparison of Tables 1 and 2, for example, shows that each of the crosses
postulated on observational grounds that has been tested experimentally has proved
inviable or sterile. Thus, free and frequent natural hybridisation cannot be invoked
to account for the variability observed in the field. On the other hand, the three
groups into which the species fall on geographical and cytogenetic grounds provide
a framework within which crossing behaviour is consistent. On this grouping, the
history of the grain amaranths may be interpreted as arising from the existence in
the wild of three progenitor species, each of which was domesticated and gave rise
to one or more cultivated grain types. The range of variation within each group is
then such as might be expected on Vavilov's law of homologous variation, and the
observed facts of contamination of cultivars by weedy types are of the kind experi-
enced with other crop plants, as for example with *Sorghum* (Doggett, 1970).

Conclusions
Some definite conclusions emerge. Firstly, *A. caudatus* and *A. edulis* are very close
genetically. Their F_1 hybrids exhibit heterosis. *A. edulis* is a grain type (cultivated
in northwest Argentine) with a recessive seed colour, whereas *A. caudatus* is predominantly an ornamental type. Their hybrid is not fully fertile and the unique
polymerous male flowers of *A. edulis* constitute a strong morphological distinction.

That gene exchange goes on between them is shown by the emergence in F_2 of plants like *A. caudatus* var *atropurpurea*. Sauer (1967) has merged the two, but while accepting their close genetic relationship, it seems preferable to maintain the taxonomic distinction (see Khoshoo, 1971).

Secondly, the three cultivar groups, *A. hypochondriacus*, *A. cruentus* and *A. caudatus—A. edulis*, represent the end points of three distinct evolutionary lines. All future work, whether it be on the geography of the group, its diversity in the field, or its improvement by plant breeding, must take account of the fact that there is no genetic interchange between them.

Thirdly, the crop—weed relationships differ in one important respect from those formulated by Sauer (1967). *A. powellii* differs in chromosome number from *A. hypochondriacus*, and the hybrid between them is sterile. On the other hand, the weed that has accompanied *A. hypochondriacus* to Asia is *A. hybridus* and not *A. powellii*, and the necessity in India to rogue out weedy introgressants as indicated by black seed, is evidence that the grouping of the $n = 16$ chromosome weed, *A. hybridus*, with the $n = 16$ chromosome grain type *A. hypochondriacus*, is correct. The likelihood that *A. caudatus—A edulis* undergoes gene exchange with *A. quitensis* may be accepted, though there has been no opportunity to test their cross compatibility.

The cross compatibility that might be expected on this grouping between *A. powellii* and *A. cruentus* has not been put to the test.

Summary

Sauer's (1950, 1967) theories of the origin and inter-relationships of the grain amaranths and their weedy relatives, are reviewed. The results of an experimental crossing programme are reported, and it is shown that genetic barriers between species are such that intercrossing is much less than was postulated by Sauer.

The four cultivated and the three weedy species are assigned on grounds of geographical distribution and of chromosome number to three groups as set out in Table 3. Crossing behaviour is such as to support this classification. Within a group, wild and cultivated species cross and give fertile offspring. Between groups, crossing may fail, or give rise to inviable or sterile hybrid plants. It is concluded that the three groups represent independent evolutionary lines, genetically so distinct as to preclude gene exchange between them.

Acknowledgments

Our thanks are due to Dr L.B. Singh, Director, National Botanic Gardens, Lucknow, for facilities and to Mr A.K. Sen Gupta and Mr T.K. Sharma for the illustrations.

Table 3   *Grouping of* Amaranthus *species on distribution and chromosome number*

Chromosome number	Ecological status	North and Central America	South America
n = 16	Weedy	*A. hybridus*	*A. quitensis*
	Cultivated	*A. hypochondriacus*	*A. caudatus*
			A. edulis
n = 17	Weedy	*A. powellii*	
	Cultivated	*A. cruentus*	

Potato

MAHESH D. UPADHYA
Central Potato Research Institute, Simla

The early European potato, although given the specific name of *Solanum tuberosum,* was shown by Salaman, on a formidable array of evidence (Salaman, 1946, 1954; Salaman and Hawkes, 1949), to be a variety of *S. andigenum.* Salaman and Hawkes argued that the northern end of the Andean region in Columbia was the area within South America whence the tetraploid potatoes were brought to Europe near the end of the sixteenth century. Studies by Simmonds (1963, 1964, 1968) showed that the modern commercial potato of the Tuberosum group, which is a long-day adapted population, evolved primarily from samples of Andigena from an area of the high Andes which he was not able to specify, and represent the product of nearly 400 years of evolution in north temperate countries. Besides adaptation to tuber formation during long days, there are morphological correlates to this evolutionary change, including reduced top growth, shorter internodes, larger leaves and reduced flowering and fruiting.

The precise time of introduction of the potato into India is not known. It is probable, however, that it was brought in either by the Portuguese who first opened the trade routes to India, or later by the British. Whether by the Portuguese or the British, it is likely that the stocks introduced were of those then grown in Europe, and not material direct from South America. The earliest reference in history is from the period from 1615 to 1619 when mention of the potato occurs in an account of the voyage of Edward Terry (1655), who was chaplain to Sir Thomas Roe, British Ambassador to the court of the Moghul Emperor Jahangir from 1615–19. Terry, in his description of Indian soil and its produce, wrote 'In the northernmost part of this empire they have a variety of pears and apples; everywhere good roots as carrot, potatoes, and others like them... are grown,' (pp. 91–2). Terry's account thus places the potato as a crop in the northernmost parts of India, probably in the hills, earlier than 1615. Similarly, Fryer's travel records (1672–81) mention the potato as a well established garden crop in Surat and Karnatak in 1675 (Watt, 1908). However, Habib (1963) in his book *The agrarian system of Mughal India (1556–1707)* is of the opinion that ordinary (now called Irish) potatoes were not among the vegetables grown then. According to him Terry's (1655) and Fryer's (1672–81) mention of 'potatoes' referred to in their travel accounts actually meant varieties of yams which were grown and formed an article of popular diet in northern and southern India during that period.

Old records of travellers who toured the Indian subcontinent include accounts of the potato as a garden crop, or under commercial cultivation in the hills as a summer crop or in the plains as a winter crop. The information available can be put in chronological order as follows:

First are the two accounts regarded as doubtful by Habib. Edward Terry (1655) had observed 'potato' in the northernmost part of India around 1615–19. In 1675 Fryer mentions the potato as a well established garden crop in Surat and Karnatak (Watt, 1908). Following these, the next report refers to 1772 when mention of the potato is made by Johnson (1847) who stated, 'A basket of potatoes... was occasionally sent by Warren Hastings to the Governor of Bombay'. Similarly Moon (1947), in his book *Warren Hastings and British India* mentions that Hastings sent Mr Bogle to Lhasa with potatoes to be given *gratis*.

Again nothing is known about the potato during the period 1785 to 1828. Captain Mundi, in his description of the Simla Hills in 1828, wrote that he saw the potato in cultivation in the gardens of many householders (Buck, 1925). Lieutenant White (1838) also recorded the potato being grown in the gardens of Simla. Captain Thomas (1846) mentioned in his *Views of Simla*, 'Potatoes we have as red and thumping as any in or out of Limerick. The finest are grown at Mahasoo, where the natives cultivate the vegetable for the Simla market and for export to the plains.' Thus apparently a red variety of potato had come to be grown as a commercial crop.

During the same period the accounts of travel by Hooker (1854) indicate that a red variety of potato was introduced in the Khasia hill region of Assam by Mr Inglis around 1830 and by 1848 the Darjeeling area was supplying seed potatoes to the plains for cultivation as a winter crop. Potato had been included as a food crop in east Nepal and Sikkim about the time of his travel to these areas in 1848. In these regions also according to Hooker, the potatoes were red in colour and were as big as a walnut. However, according to the Gazetteer of Assam (1906, pp. 73–4), which deals with Khasia, Jaintia, Garo and Lushai Hills, the potato crop was first introduced in 1830 by Mr David Scott (Watt, 1908).

In the Nilgiri Hills, the potato was being grown as a regular crop around 1860, according to Markham (1862).

The historical account given above provides ample evidence that by early in the nineteenth century the potato had established itself as an important vegetable crop in the hills and plains of India. Thus, all the introductions of the potato were from stocks existing before the great blight pandemic of the 1840s which virtually wiped out the European varieties of that day. This view receives support from the fact of the Indian potato varieties sent by Pal and Pushkarnath (1951) to Dr Salaman in England for identification, a large proportion proved to be very old varieties which had gone out of cultivation in that country.

Detailed studies of Indian potato varieties by Pal and Pushkarnath (1951) led them to recognise fifteen distinct locally established, or *desi*, varieties. The range of diversity among them may be ascribed to three sources, diversity in the original introductions, segregation in seedling offspring of heterozygous clones or new hybrids in India, and somatic mutation. The first is likely to have been important, especially as Salaman was able to relate Pal and Pushkarnath's material to old European varieties. A few of the *desi* varieties (e.g. Phulwa) set seed in India, so some naturally occurring seedlings may have survived. Nevertheless, in most varieties seed setting is rare, and it is likely that any addition to the variability in India has come predominantly from somatic mutation.

The *desi* varieties have been said to resemble the Andigena group (Swaminathan,

Table 1   *Characters used in scoring potato varieties for determining relative affinity to S. andigena*

Character	Grading points (numerical rating)		
	0	1	2
A (1) Number of stems	Many	Moderate	Few
(2) Thickness of stem	Thin	Medium	Thick
(3) Haulm colour	Strong	Medium	Very faint or nil
B (4) Leaf setting	Stiff	Moderately drooping	Drooping
(5) Leaf structure	Open	Medium	Close
(6) Leaflet shape	Lanceolate	Lanceolate—ovate	Ovate to oval
(7) Leaflet size	Small	Intermediate	Large
C (8) Stolon length	Long	Intermediate	Short
(9) Tuber shape	Cylindrical	Ovoid	Globose
(10) Eyes	Very deep—deep	Medium	Fleet
(11) Flesh colour	Strong yellow	Medium	White

Table 2   *Affinity index values of Indian* desi *varieties and three of the varieties evolved in India*

Serial numbers	Varieties	Traits			Total
		Stem	Leaf	Stolon and tuber	
	Desi *varieties*				
1	Silbilati	1	0	2	3
2	Coonoor White	4	1	3	8
3	Shan	4	4	1	9
4	Red Long Kidney	4	4	2	10
5	Dessa	3	4	3	10
6	Darjeeling Red Round	4	3	3	10
7	Chamba Red	3	3	5	11
8	Coonoor Red	3	4	4	11
9	Agra Red	3	3	5	11
10	Phulwa	4	5	2	11
11	Sathoo	4	4	4	12
12	Dhantauri	3	4	5	12
13	Gola type A	4	3	5	12
14	Gola type C	5	5	5	15
15	Gola type B	6	5	5	16
	Newly evolved *varieties*				
16	Kufri Sindhuri	4	2	5	11
17	Kufri Safed	4	3	5	12
18	Kufri Khasigaro	3	6	6	15
	Maximum score	6	8	8	22

1958). Sinha and Pushkarnath (1964) undertook a detailed study of them to establish their affinity with Andigena, taking into account fourteen distinguishing characters from the descriptions of Pal and Pushkarnath (1951). They used Anderson's (1949) 'hybrid index' technique to calculate an affinity index for each variety. For the present study, Simmonds' (1963) work was taken into account. He showed that in distinguishing the Andigena and Tuberosum groups of potatoes, Salaman's (1946) leaf index was unreliable. The list of characters used by Sinha and Pushkarnath (1964) to calculate the affinity index was therefore modified, and the number reduced to eleven. These are set out, with a three-grade scoring system for each character, in Table 1. The scoring system used ranks

near Andigena	0
intermediate	1
near Tuberosum	2

It is arbitrary and subjective, but it affords a means of compiling a single index giving a measure of the resemblance of a clone to one or other of the two important potato groups.

Scores for the three main groups of characters, and summed affinity ratings, are given in Table 2 for fifteen *desi* varieties described by Pal and Pushkarnath (1951), and for three newly evolved varieties. It will be seen that of the *desi* varieties, Silbilati (syn. Helora) ranks very close to Andigena. The others are not close to Andigena, but form a rather compact group in the intermediate range, from 8 to 15. The first two of the three new Kufri varieties are selections from *desi* X Tuberosum hybrids. All three are intermediate, but rank towards the higher end of the distribution of the *desi* varieties.

The potato introductions in India were grown and selected under environments consisting of two opposite agroclimates. In the hills they were grown at high altitude and under long summer days, and in the plains at low altitudes and under short winter days. Therefore, the selection pressures of the Indian environment are likely to be different from those either of high-altitude low-latitude America or of low-altitude high-latitude Europe. The grouping of characters that emerged during the selection of Tuberosum in Europe from the Andigena of South America may not hold in the Indian situation, and this must now be examined. The data in Table 2 are plotted as a pictorialised scatter diagram (after Anderson) in Fig. 1. It is clear from Table 2 and Fig. 1 that the correlations between the three main groups of characters are low, suggesting that though there has been considerable departure from the Andigena characteristics, there has been no tendency in India for a Tuberosum type to emerge as a result of selection over the past three centuries. Further, there is little tendency for Tuberosum types to be outstanding in the breeding programmes of the recent past. Indeed, even in crosses not involving *desi* types, the selection pressure is such that the clones selected show intermediate values, the selected types resembling Andigena in the major foliar characteristics, and Tuberosum in most of the tuber characters.

Two of the most important characters distinguishing Tuberosum from Andigena, leaf size and day-length response, are not included in those used in scoring for the affinity index. As an index of leaf size Simmonds (1964) used his index *d*, the

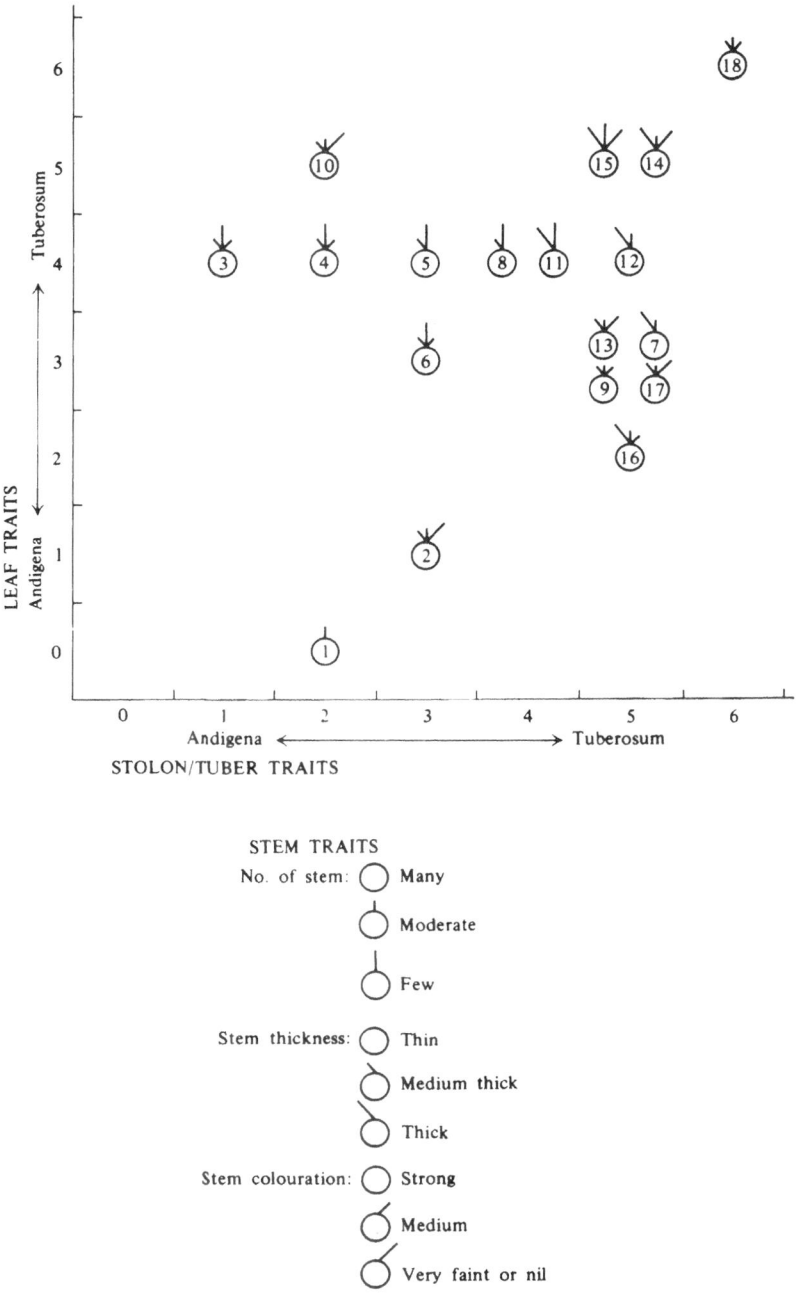

Fig. 1 Pictorialised scatter diagram showing association of characters in Indian potato varieties (refer to Table 2 for the names of the varieties denoted by numbers).

Table 3   *Frequency arrays of d-values of clones in various populations*

	Total no. of clones.	Frequency classes of d-values				
		41–60	61–80	81–100	101–120	121–140
Andigena collections	32	19	11	2	0	0
Kufri Safed self lines	45	5	25	14	1	0
F_1 plants from crosses between two Andigena clones: 950 × 1254	29	5	18	6	0	0
1250 × 950	29	7	14	8	0	0
Darjeeling Red Round clones	35	4	30	1	0	0
Phulwa clones	39	37	2	0	0	0

Table 4   *Affinity index and* d *measurements of some potato varieties and* andigena *collections*

Variety	Affinity index	d-values (Mean)
Desi *varieties*		
Darjeeling Red Round	10	69
Phulwa	11	52
Newly evolved varieties		
Kufri Safed	12	76
Kufri Sindhuri	11	139
Kufri Khasigaro	15	113
S. andigena *collections*		
CP 1736	10	71
CP 1738	12	80

distance in millimeters between the tip of the terminal leaflet and the point of insertion of the first pair of laterals of one of the largest leaves borne about half way up a flowering shoot. The d-index was measured on an Andigena collection (32 clones), on selfed lines of Kufri Safed (45 lines), on F_1 plants from crosses between two pairs of Andigena clones (29 clones of each), and on a collection of clones, from as wide an area as possible, of two *desi* varieties, Darjeeling Red Round (35 clones) and Phulwa (39 clones). The material was grown under the short-day conditions of the winter months in the plains. Measurements of d-values were taken from the largest leaf from at least ten plants of each clone to give a reliable mean. The data are presented as frequencies in different d-value classes in Table 3 and plotted in Fig. 2 as percentages of individuals of a clone in each class. Among Andigena collections the greatest percentage of individuals is in the lowest d-value class. The clones of Phulwa also show almost the same distribution, indicating that the variety Phulwa is in this character close to Andigena. The clones of the rest of the populations studied show a normal distribution, with modal values higher than in the Andigena collection, an indication of increasing leaf size. The similarity of the

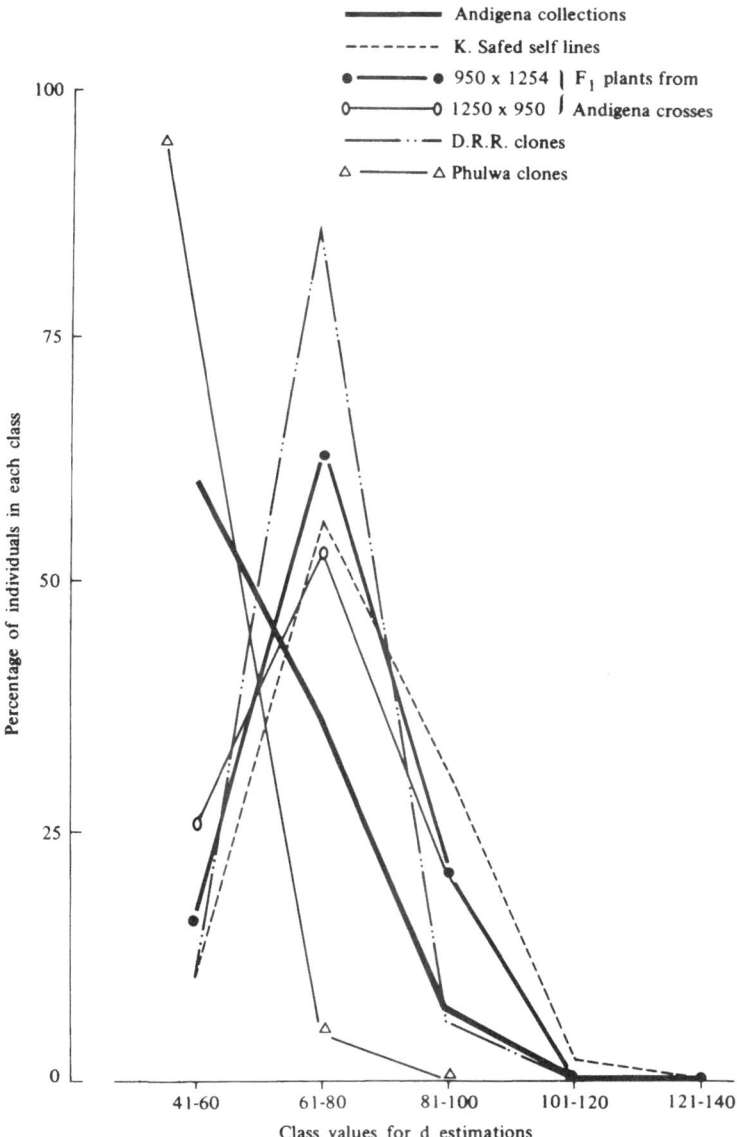

Fig. 2. Frequency distributions of *d*-values, expressed as percentages of the total number of clones in different populations.

Table 5

Table 5   *Effect of photoperiod on d-values and tuberisation in some potato varieties*

Variety	d-values			Days to tuber initiation	
	S.D.	L.D.	Per cent reduction under L.D.	S.D.	L.D.
'OT'	88	80	9	37	55
EM/B-1399	100	80	20	37	No tuberisation
Kufri Sindhuri	120	92	23	36	No tuberisation
Kufri Shakti	124	90	27	36	No tuberisation

S.D. = Natural daylight period of 8 hours alternating with 16 hours of dark period.

L.D. = 24 hours continuous light (natural daylight plus artificial illumination).

d = Values taken after 8 weeks of growth.

F_1s from Andigena crosses to *desi* types shows how rapidly the *desi* leaf character is established in breeding material under Indian conditions.

Affinity index is compared with d-value for seven varieties in Table 4. In d-value the *desi* varieties rank with Andigena on these data. Of the new varieties, Kufri Sindhuri with the highest d-value (typical of the Tuberosum group) has an affinity index similar to the *desi* varieties, whereas Kufri Khasigaro, having a lower d-value than Kufri Sindhuri, has a higher affinity index.

The second of the important distinguishing characters among the two groups is the adaptation of the Tuberosum group to tuberisation during long days, while the Andigena group only forms tubers during short days (Simmonds, 1963, 1964). Studies were initiated at the Central Potato Research Institute to classify the photoperiodic response of *desi* varieties and of the varieties bred at the Institute. The assessment of varieties with respect to day-length reaction should be based on a consideration of critical day length and the time of tuber initiation under two extreme photoperiodic conditions. All potato varieties of the Andigena and Tuberosum groups have been found to be short-day types with respect to tuber initiation. The varieties with strict short-day requirements are qualitatively distinct, and will not tuberise under continuous light; the quantitative short-day types have the potential to tuberise under both extremes of photoperiods, with a difference in the time of tuber initiation. The day-neutral types have the potential to tuberise within the same period under both extremes of photoperiod (Upadhya, Purohit and Sharda, 1972). Data for four varieties on the relationship of photoperiodic response with respect to tuberisation (from Purohit, 1970) and leaf size as expressed by d-values, are presented in Table 5.

All four varieties showed a photoperiodic response. The variety OT, being a quantitative short-day type, formed tubers under 24-hour illumination, but tuber initiation was delayed by 18 days. In the other three varieties, which have a qualitative short-day response, tuberisation was not initiated at all under 24-hour illumination. Three of the four varieties had high d-values, and d-values also exhibited a photoperiodic response, being lower under 24-hour illumination than in short days. Moreover the response in d-value was greater in the varieties that failed

Table 6 *Morphological characters of the Indian* desi *race*

	Characters	Features of *desi*
A	1. Number of stems	Moderate to few
	2. Thickness of stem	Medium
	3. Haulm colour	Medium to very faint
B	4. Leaf setting	Stiff to moderately drooping
	5. Leaf structure	Open
	6. Leaf shape	Ovate
	7. Leaf size	Intermediate
C	8. Stolon length	Intermediate
	9. Tuber shape	Ovoid
	10. Eyes	Deep to medium
	11. Flesh colour	Medium yellow

to form tubers under 24-hour illumination. Finally, the three varieties with d-values in the Tuberosum range were those that showed the Andigena character of failure to form tubers under 24-hour illumination. It is very significant to note that qualitative short-day varieties, being highly day-sensitive, show a greater degree of change in the d-values, whereas the quantitative types, being less sensitive to photoperiod, show a lesser degree of change in the d-value. It has now been found that the true day-neutral types show a very low order of change in d-values (Upadhya and Purohit, unpublished data).

The conclusion reached above that Indian conditions have not been conductive to the emergence of the Tuberosum group of characters is thus reinforced. The selection pressures acting on the limited diversity of the early potato populations of India have not tended to create a new Tuberosum. There has arisen a distinctively Indian type, still recognisably close to the Andigena material but diverging from it in characters that affect productivity in the climatic regime of northern India. A new Indian *desi* race has emerged. The morphological characteristics of this race for the eleven characters used for determination of the affinity index are given in Table 6. The d-value ranges between 50 and 80, whereas the photoperiodic response is qualitative short-day type with a critical day length around 12 hours (Purohit and Upadhya, unpublished data).

The breeding programme which was started in the early 50s, has not yielded results of significant value. This has been mainly because the physiological—genetic requirements for success were neither well understood nor sufficiently clearly defined to enable the breeder to evolve suitable selection criteria. The breeder has found that the characters with which he has been concerned are not assorted in the Indian material in the way he would have expected from European experience. Yet his breeding policy has been based, not so much on a pattern of characters most desirable in the light of evolutionary influences met within India, as on the character combinations evolved in other countries. Moreover, even the combination of characters which had become established in the acclimatised *desi* varieties under past agroecological pressures is not optimal for present and future needs in the currently changing agricultural pattern.

In the Indian subcontinent the potato is grown as a summer crop in the hills, as a rainy and spring crop in the plateau region, and as a winter crop in the plains. Therefore, the agroclimatic requirements of different regions necessitate a breeding programme directed towards developing different varieties suitable for different regions. The extensive studies on the photoperiodic responses of potato varieties conducted at this Institute have made it possible to define the needs in a potato variety to fit into the present-day intensive agriculture pattern (Upadhya *et al.* 1972). It is now well understood that for a variety to be widely adaptable it has to (i) be insensitive to day length and temperature with regard to tuber initiation and bulking, (ii) have a high photosynthetic efficiency and low photorespiration, and (iii) have a short duration so as to fit well with intensive cropping patterns.

As a first attempt, therefore, as a blueprint for Indian potato breeding the following ideotype is suggested:

(i) Day-length neutral with regard to tuber initiation;
(ii) Insensitive to photothermoperiod in the bulking phase;
(iii) High photosynthetic efficiency and low photorespiration to give a high net photosynthetic yield;
(iv) Short duration, preferably of 75–90 days; and
(v) Very high degree of field resistance/immunity to late blight, brown rot and virus X.

In working towards this ideotype a wide range of Andigena and Tuberosum material is used and is crossed with *desi* varieties. It is expected that a variety having these characteristics will give the highest yields in terms of dry matter/unit area/unit time. Current work, though in the initial stages, confirms the theoretical expectations outlined above. Successful breeding on these lines may be expected to alter our conception of the *desi* type, but the type will remain characteristic and distinct, and adapted to Indian conditions.

Acknowledgments

I am grateful to Messrs Maria Bros., The Mall, Simla, for having allowed me to look through and take notes from the rare books in their collection.

5
Review

Crop plant evolution in the Indian subcontinent

J.B. HUTCHINSON

St John's College, Cambridge

The agricultural origins of crop plants

One of the greatest advances in evolutionary studies of crop plants in recent years has been the excavation and dating of some of the earliest agricultural settlements. Evolution under domestication cannot be older than agriculture, and the dating of agricultural beginnings in various parts of the world provides a time scale against which the development of crop plants can be studied. In India, there is still much to be learnt about the beginnings of agriculture. Remains from simple farming peoples in Kashmir and Baluchistan give indications of the nature of early farming patterns, but they are outside the area with which we are concerned, and indeed they are no older than the mature agriculture of the Harappans in the Indus Valley and western India. The Harappan culture has been extensively studied and is dated with confidence to the period between 2300 and 1750 B.C., but since there is extensive evidence of both an advanced agriculture and a sophisticated textile craft, the Harappan evidently does not represent the beginnings of Indian agriculture. Indeed, Thapar (1972) has excavated at Kalibangan in Rajasthan, a pre-Harappan ploughed field. The plough pattern revealed there is the same as that practised in present-day mixed cropping in the *rabi* season. Thus the earliest agriculture so far excavated in India was one in which cultivation practices which persist to this day had already been established.

Enough is known, however, of the range of crops of the Harappans, and of the appearance of new crops later in the archaeological record, to date the evolutionary process in Indian crops within reasonable limits. The crops of the Harappan period were chiefly of west Asian origin. They included wheat, barley and peas. Of indigenous Indian origin were rice, cotton and probably sesame. Rice first appears in Gujarat and Bihar, not in the centre of the Harappan culture in the Indus Valley. There is some, rather doubtful, evidence that African crops were also grown by the Harappans. There is a record of *Sorghum* (jowar) from Sind and of *Pennisetum* (bajra) from Gujarat. Sesame, recorded from the Punjab, is of uncertain origin, having been ascribed by some to Africa and by some to India. The earliest record of the third African cereal, *Eleusine* (ragi) is from Mysore, about 1800 B.C.

Southeast Asian crops of importance in India are sugarcane and bananas, and they are not of the type likely to leave identifiable remains in the archaeological record. Consequently, nothing is known of them before historic times, but both appear in the early literary record.

Crops of American origin include maize, grain amaranths, and potatoes. The dating of the introduction of maize is uncertain, the characteristics and distribution of some forms being such as to lend support to the view that they reached India in pre-Columbian times. The potato, on the other hand, was certainly a late introduction by either the Portuguese or the British.

This is in outline the history of crop establishment in India. The oldest crops have been under the selective pressures of the Indian environment for about 4500 years. Probably all save those of American origin have been so exposed for at least 2000 years. The post-Columbian introductions from America, on the other hand, have been in India less than 500 years.

The source of a cultivated species affects very greatly the genetic situation under which it responds to selection. Considering the beginnings of domestication, the adoption of a species by man and sowing and harvesting under controlled conditions do not isolate the incipient cultivar from its wild relatives. Many crop plants are self-fertilised, or even vegetatively propagated, but all have floral morphology adapted for cross-pollination, and there is evidence (Hutchinson, 1965) that in the earlier stages of domestication they were cross-pollinated. Hence the early stages of domestication must be regarded as a period when the cultivar was under selection to meet the needs of agricultural man while it remained in genetic contact with wild-growing relatives subject to the selection pressures of the natural environment. This 'disruptive selection' situation has been studied in detail by Thoday (1964), who has shown that response is such as to establish distinct forms, even though gene exchange continues. The presence of wild weedy relatives with which there is continuing gene exchange is important evidence of the place of origin of a crop plant. Mehra (1963b), for example, has shown that the wild and weedy *Eleusine* of African finger millet fields is of the same ploidy as, and is genetically compatible with, the cultivated tetraploid species *E. coracana*, whereas the corresponding *Eleusine* weed of Indian finger millet (ragi) fields is diploid. He concludes that the crop plant is much more closely related to the African weedy species than to the Indian. Alongside Mehra's botanical evidence of the affinity between *Eleusine coracana* and its African relative *E. africana* must be set Vishnu-Mittre's evidence that in deposits dated 1800 B.C. at Hallur, grains matching those of *E. coracana* and *E. indica* were found associated together.

The genetics of domestication

Gene exchange may be terminated in various ways: by polyploidy, as in the wheats; by the development of genetic isolation barriers, as in the early- or late-flowering *Brassica campestris* (Sarson) cf. Murty *et al.* (1972), or by the establishment of the crop in areas beyond the range of the wild relatives. There remain, however, many species in which there is continuing contact between the cultivar and its wild relatives, and the persistence of gene exchange has substantial consequences for the wild type, and probably also for the cultivar.

The effects of gene exchange under disruptive selection have been discussed by Doggett (1970) in his account of the evolution of the cultivated sorghums. He has shown that the sorghums evolved in Africa from wild grass species under human selection. Gene exchange with the wild species has gone to the present day, though the increasing divergence between wild and cultivated, the increasing areas of cultivation, and the spread of the crop beyond the natural limits of the wild species have together reduced the opportunities for hybridisation. However, the impact of agriculture on the natural vegetation has opened up a new habitat for weedy plant species. Under the shifting cultivation so widely practised in Africa, there are large areas of abandoned land in process of reversion to natural vegetation, alongside the farmers' fields. In these, grassy sorghums intermediate between wild and cultivated types are able to establish themselves.

These weedy types are in genetic contact with the cultivated types, and in characters that are neutral with respect to the weedy habitat they come by gene exchange to resemble the local cultivars. Thus arose the situation that over the great diversity of the African sorghums, the weedy types and the cultivars in any area are similar. This is so to the extent that Snowden (1936) concluded that several different wild sorghums had given rise to different cultivated species. Doggett adds that, not only are the genotypes of the weedy sorghums affected by those of their cultivated neighbours, but the introduction of a new cultivar into an area is followed by a genetic change in the direction of the old type of the area through gene exchange with the local weedy forms.

Doggett has shown that the generation of weedy forms, and the impact of these on the cultivars, has gone on in the United States, where grassy species of *Sorghum* are grown for fodder. He suggests that in China also, where the diploid *S. propinquum* is native, genetic material from *S. propinquum* has contributed to the development of the *kaoliang* sorghums. In India, however, the wild sorghums are tetraploid and they have had no genetic influence on the grain sorghums, which are diploids.

Comparison of the situation in India and Africa is pertinent to another feature of disruptive selection. Thoday (1964) in his analysis of the effects of such selection in *Drosophila,* showed that genetic variability was thereby increased, and Doggett (1970) has supposed that disruptive selection has contributed to the diversity of both cultivars and weeds in Africa. Be this as it may, the diversity of sorghum cultivars in India where there is no weedy type, is not noticeably less than in Africa where the weedy types accompany the cultivars over most of the continent.

The gene exchange situation between cultivar and wild ancestor, the consequences of disruptive selection and the emergence of weedy forms, have been set out in some detail for sorghum because they have been so admirably elucidated by Doggett. Sorghum serves, therefore, to introduce the principles involved, and the extent to which they apply to other Indian crop plants will now be discussed.

The situation is determined primarily by the distribution of the wild relatives of the cultivated species. The crop plants of west Asian, African and American origin are in India geographically isolated from their wild relatives. Those of Indian origin are still in contact with their immediate relatives, and those of southeast Asia have in India come together with species sufficiently closely related to have contributed to the cultivars by hybridisation. Indeed, in regard to crops of Indian or African origin, one of the important pieces of evidence on the continent of origin is the distribution of wild relatives, and of areas of intercrossing between wild species and cultivars. For example, Mehra's evidence on the weedy relatives of *Eleusine coracana* quoted above, indicates an African origin for the crop, and De's evidence (pp. 81—7) on the relations between *Cajanus* and *Atylosia* demonstrates an Indian origin for *Cajanus.*

Of the indigenous Indian crops, rice and cotton may be considered. In rice, the situation is very similar to that in sorghum. The truly wild perennial forms are generally regarded as representing the type from which the cultivated species was derived. Wild annual forms are common and persistent weeds in rice fields and in neighbouring habitats. Intercrossing between the weedy types and the cultivars is common, and in long-established crop varieties, weedy types have arisen that resemble the associated cultivar to an extent that makes elimination by weeding very difficult. Crossing between the wild perennial and the cultivated rices also goes on, but probably less commonly, and contributes to the generation of further annual weedy types. The contribution to diversity among wild, weedy, and cultivated types is described earlier (pp. 58–60) by Shastry and Sharma.

The situation in cotton is rather different. The indigenous cottons of the Old World are diploid, and in these the cultivars may be expected to give fertile hybrids with their wild relatives. The cottons of the New World, which are now extensively grown in India, are tetraploid and are separated from the Old World species by a virtually complete sterility barrier. Even within the Old World species, crossing between wild forms and cultivars is not common, chiefly because environmental circumstances are less favourable, both to the survival of wild types and to the establishment of weedy forms. All species of *Gossypium* appear to be very palatable to grazing animals, and with increasing stock populations, both wild and semi-wild cottons tend to be grazed out.

Wild members of the diploid *G. herbaceum* are known in southern Africa beyond the range of intercrossing with commercial diploids. Wild forms that probably do intercross with cultivars occur in southern Arabia and along the coast of Baluchistan and Pakistan as far as Karachi. Wild forms of *G. arboreum* that must be regarded as in genetic contact with cultivars of the same species have been recorded extensively in western India (see Santhanam and Hutchinson, pp. 89–90). All these forms are perennial, and the transition among the cultivars from perennial to annual forms in the last 500 years has not been matched by a parallel evolution of annual weedy forms. That such annual weedy forms can arise is beyond dispute. In Egypt the 'Hindi weed' cotton which contaminated the long staple crop for many years, was of this type. 'Hindi weed' is an exception, and there must be reasons why annual weedy cottons were so rarely established when they are so common and so successful in sorghum and rice.

In the cereals, the shattering panicle is a dispersal mechanism that operates at or before harvest, and much of the seed of the weed falls to the ground and escapes the hand of the harvester. In cotton, the lint-covered seed remains in the capsule for some time after the capsule opens, and hence is as readily harvested from the wild form as from the cultivated. Moreover, seed cotton must be processed to separate the lint from the seed, and this gave an opportunity, even with simple gins, to exercise quality control over the seed supply. This facility has been exploited to the extent that for half a century the maintenance of varietal purity was more advanced in cotton than in any other tropical crop. There can be little doubt that the commercial pressures that led to the establishment of the annual and to the maintenance of quality standards prevented the evolution of annual weedy forms.

The relations between the southeast Asian crops, sugarcane and bananas, and their Indian relatives are of a different kind. Both crops have been established in Indian agriculture since before the dawn of history, but neither leaves evidence in the archaeological record whereby its full antiquity might be established. Both are vegetatively propagated, and changes under domestication have taken place in a context of wide crosses, polyploidy, and the development of sterility.

The genesis of the north Indian sugarcanes (the *Saccharum barberi*/*S. sinense* group) has been discussed by Parthasarathy (1947), Bremmer (1966) and Price (1968). Taking Barber's (1915, 1917) original classification into five groups on morphological grounds, first Bremmer and then Price refined it, and showed that the morphological classification was confirmed by the pattern of chromosome numbers. Parthasarathy propounded the idea that the north Indian sugarcanes arose through hybridisation between *Saccharum officinarum* and *S. spontaneum*. That the wild north Indian grass, *S. spontaneum*, has contributed its genotype to the north Indian sugarcanes is now generally agreed. That the southeast Asian *S. officinarum* was also involved in the original hybrids is not so readily accepted. Bremmer (1966) postulated a sugarcane with a basic chromosome number of 17 (*S. officinarum* has 2n = 80; Stevenson, 1965) and drew attention to Parthasarathy's record of an Indian sugarcane with 2n = 68. Bremmer, on this basis, compounded the chromosome numbers of his north Indian sugarcanes from multiples of 17 from the female parent, plus 48 or 56 from *S. spontaneum* as male. Price (1968), however, showed that chromosome numbers in the group are not regular as suggested by Bremmer's data. In his view the range of numbers observed is as might be expected if the clones arose by intercrossing aneuploid parental clones of interspecific hybrid origin. He concluded that 'The nature just propounded for the *S. sinense* groups limits the need for proposing elaborate mechanisms to explain their origin' and 'the general hypothesis of Parthasarathy suffices'.

Parthasarathy gives reasons for believing that *S. officinarum* sugarcanes have been in cultivation in peninsular India from early times, and cites Dravidian literature of the Sangam period before the second century A.D. He supposes that the hybridisation with the wild indigenous *S. spontaneum* that gave rise to the north Indian sugarcanes took place in the area of West Bengal, Bihar and Orissa. The emergence of the north Indian group of clones led to the establishment of sugarcane as an important crop in an area beyond the tropical belt to which *S. officinarum* is adapted. This constitutes a major evolutionary advance in the crop. The evolution in India of an introduced crop has been dominated by the contribution to it of the genotype of an indigenous wild relative. Moreover, advantage has been taken of this potential in modern sugarcane breeding. *S. spontaneum* has become a major source of disease resistance and of hardiness for sugarcane breeders throughout the tropics.

A parallel situation exists in the bananas. Simmonds (1959, 1962) has given an account of the origin and spread of the edible clones of bananas. He has shown that they arose by the establishment of parthenocarpy and sterility in *Musa acuminata* of Malaysia and Indonesia. *M. acuminata* carries the A genome and the basic cultivars are diploid (AA) and triploid (AAA). The spread of the *M. acuminata* cultivars brought the crop into contact with the related *M. balbisiana*, which carries the genome designated B. India was one of the areas into which *M. acuminata* spread and this contact occurred, and there arose in India clones with the genomic constitution AB (diploid), AAB and ABB (triploids). This group of clones is of sufficient importance for Simmonds (1962) to regard India as a secondary centre of origin of cultivated bananas. Thus in bananas, as in sugarcane, the crop has evolved in India by the addition of the genome of an indigenous wild species to that of an introduced cultivar.

Sugarcane and bananas may be compared with potatoes, a vegetatively propagated crop of American origin, more recently introduced into India. The crop has an Indian history of the order of 200 years only. The Indian acclimatised stocks show clear evidence of the nature of the material originally introduced, and there has been nothing contributed to their genotype from indigenous Indian relatives. Nevertheless, the selection pressures of the Indian agricultural environment have endowed India with a group of clones substantially different both from the original parental material and from the clones emerging from similar basic stocks under the selection pressures and the breeding policies of western European agriculture.

The rate and extent of evolutionary change that has gone on in India varies not only with the length of time the crop has been in the country, as in the comparison between sugarcane and bananas on the one hand and potatoes on the other. Wheat and barley are both to be found in the earliest remains of farming peoples so far excavated, and in some early agricultural periods barley was the major crop (Bakshi and Rana, p. 47). Yet the diversification of wheat in India has been of an altogether higher order than that of barley. Comparison with the history of the two crops in Europe shows that this is not because the potential in barley is low. It must rather be ascribed to the acceptance of wheat as a superior food grain, and the consequent disproportionate increase in the size of the wheat crop, and in the attention devoted to it. Of the crops discussed in this volume, the position of *Coix* is similar to that of barley. Koul (p. 65) concludes that genetic potential is available in the crop if it were ever desired to develop its agricultural status, but over the period of Indian agricultural history the success of the domestication of rice has resulted in a degree of dominance of the rice crop that has precluded the expansion and improvement of *Coix*. By contrast, the diversity of castor, for long a crop of minor agricultural interest, has been increased by its adoption as an ornamental in horticulture. Thus, at a time when its agricultural importance is increasing, the range of material available to the plant breeder is greater than is to be found in *Coix* or — at least as far as India is concerned — in barley.

Genetic diversity

Genetic diversity is the raw material of the plant breeder's craft, and the sources of variability for future crop improvement must now be considered. Improvement begins with the exploitation of the variability within natural crop populations. This is very rapidly exhausted in vegetatively propagated plants, rather less rapidly in self-fertilised seed propagated crops, and only slowly in out-pollinated seed-propagated crops. Diversity between crop populations comes next in order, and immediately the plant breeder is faced with the opportunities, and the problems, of exploiting genetic material from other countries beside his own.

Going outside the strict confines of the crop itself, there is the variability contributed by weedy relatives, often closely genetically related to nearby crop populations, and by truly wild relatives, some of which may cross freely with the cultivar, and some of which may be brought into the genetic fold only with difficulty. Riley and Kimber (1966) have used the term 'comparium' for the group of taxa linked to a crop plant by the capacity to form viable hybrids with it.

The crop plants of west Asian, African and American origin are in India geographically isolated from their wild relatives. No gene exchange outside the cultivars is possible, and evolutionary progress has depended on the exploitation of the variability originally introduced or subsequently generated by mutation. Only in recent times has the gene pool been augmented by the deliberate introduction of material from other countries, in the interests of efficiency in breeding programmes.

It is on account of the needs of the breeders of crops grown beyond the range of the original cultivars and their wild relatives that there has been in recent years so much interest in the collection and maintenance of material from centres of origin and of early cultivation of crop plant species. The emphasis has been particularly on the more primitive forms of the cultivar, still persisting in areas unaffected by modern agricultural techniques. It is felt that this material constitutes a particularly valuable component of the available variation, and one most likely to be lost through the adoption of modern varieties.

While this point of view has validity, it can be pressed too far. Among cotton breeders, for example, in the decade following the Second World War the primitive cultivated cottons of central America were extensively collected. They were all perennials, quality was low, and no important resistances to either pests or diseases were identified. Indeed, their interest was academic only. The advances in cotton breeding in recent years have come from exploiting the diversity among advanced forms of the cultivated species, which have differentiated under different climatic and agricultural conditions. Thus in the New World cottons, collections of advanced breeding stocks from the United States, Russia, India and Africa provide the resources of the plant breeder, rather than the primitive perennials of house yards and subsistence holdings of Mexico and Guatemala.

In wheat, on the other hand, the recent great advances in varietal productivity have come to some extent from wide crosses with distantly related species – cf. the resistance of Hope to stem rust, derived from Yaroslav emmer – but very largely from recombination in the diversity existing among wheat cultivars from a wide range of temperate and subtropical countries. The breeding of the dwarf wheats of the 'Green Revolution' involved dwarfing genes from Japan transferred first to a North American winter wheat background, and thence to short-term low-latitude varieties from Mexico, Kenya and Australia. Recent breeding work in India has added a genetic contribution from south Asia, but in none of this material have the stocks of the west Asia centre of origin been included.

The value of collections from peasant cultivators in the original home of a species is not always so low as in cotton. South American collections of potatoes of the Andigena race, for example, figure largely in the breeding programmes of all potato-growing countries, including India.

Since genetic diversity is the source of all breeding improvement, the collection and preservation of variable material of all kinds is vital for the maintenance of the basic breeding resource. Nevertheless, much of the primitive material is outdated, and attention should be given more to the various forms of adaptation to advanced agricultural circumstances, than to the relict material of simpler farming systems. Examples referred to in this volume are natural and induced dwarf mutants in barley, spineless forms in castor, and brachytic mutants in maize. Further, the resources available in what Riley and Kimber (1966) have called the 'comparium' of a crop plant are also worthy of the attention of the collector. Riley and Kimber's (1966) discussion of the transfer of alien variation to wheat sets out the problems and the potential of crop improvement by gene transfer across wide crosses. This potential is particularly high in vegetatively propagated crops, as can be seen from the impact of species crossing on the generation of Indian cultivars of sugarcane and bananas.

The concept of the species

For wise choice of material for inclusion in a breeding programme, a plant breeder needs to know the nature and extent of the partition of genetic diversity within the 'comparium'. For this, the standard taxonomic classification is not sufficient since it gives no adequate information on barriers to gene exchange within the group. Two examples may be given. Rao's account of the ploidy relations of the *Solanum nigrum* group establishes the existence of cytogenetic isolation barriers. On the other hand De's (pp.84−5) account of hybridisation of *Cajanus* shows conclusively that what is known taxomically as the separate genus *Atylosia,* is in fact sufficiently closely related to *Cajanus* to allow easy crossing between *Cajanus* and three species of *Atylosia,* giving fertile F_1s and subsequent generations.

Whether or not there is a genetic basis to the taxonomist's concept of a species has been debated long and inconclusively. It must be accepted that a species can only be defined in terms of the available information, and the taxonomist's task is to devise a classification on what is available, and not on what one would wish ideally to know. Crop plant species generate an interest far beyond that devoted to the flora as a whole, and from the information thus gained it is possible in these species to make a classification that is based on an understanding of the genetic nature of the differences between taxa.

The genetic analysis of species differences was established on a firm footing by Harland (1936). Harland's concept of a species was of a population sharing a common, internally balanced, genotype. Diversity within that genotype may be great, but in hybrid programmes within that population, the individuals will be balanced, with normal vigour and reproductive ability. Hybridisation between members of populations differing in their genotype will yield a proportion of unbalanced, unthrifty and sterile offspring. As the difference between the genotypes increases, so will difficulties in crossing arise, and sterility will become common.

Harland's system works well in cotton, and has provided a sound basis both for classification and for genetics and breeding, for nearly forty years. It provides a sound genetic base for the study of other seed-propagated species. Two examples from this volume may be cited, rice and the grain amaranths. In rice, Shastry and

Sharma (pp. 57–8) have drawn attention to the compatibility between cultivated rices and their weedy relatives in the same area on the one hand, and the differentiation between geographic races (including both cultivars and weedy forms) on the other. Thus in Harland's terms, differentiation between the *indica* and the *japonica* rices has gone on to the extent that there is substantial genetic breakdown in crosses between them. It seems likely that the west African rices of the *glaberrima* group constitute a third distinct genotype. The contrast between free crossing between cultivar and wild in the same region, and genetic differentiation between geographically distinct groups, suggests that the geographic distinctions are the older, and the impact of domestication relatively recent. It thus may well be that the geographical and genetic differentiation evident in the cultivated forms stems at least in part from previously existing differentiation in the wild ancestor. It follows that in the vast tract of tropical country from west Africa through India and the Philippines to south China, rice was probably domesticated in many separate places from local forms of a pantropical wild progenitor, in which there was already a degree of genetic differentiation.

In the grain amaranths Pal and Khoshoo (pp. 132–5) have shown that there is a more complete genetic differentiation in the group of species from which the cultivars have been derived. By taking account of the results of hybridisation experiments, the relations of the cultivated species with their wild and weedy relatives can be shown to differ substantially from those deduced solely from morphological and geographical considerations.

Recent work has shown in many crop plants a closer relationship between cultivars and wild and weedy types. With the knowledge now available of the brief period over which man has been practising agriculture, it is not surprising that so many crop species can now be seen to be conspecific in the Harland sense with their nearest wild relatives (Hutchinson, 1971*b*).

Concepts of species relationships are complicated by polyploidy, vegetative reproduction, and sterility. Polyploidy may create a barrier that is virtually complete, even between close relatives. Vegetative propagation makes it possible to build up large stable populations of hybrid stocks so diverse in parentage as to have no prospect of survival by sexual reproduction. When sterility is added, as in the extreme case of the bananas, each clone has a genetic isolation as great as that of a seed propagated species.

It will be apparent that the extent to which the genetic diversity within the 'comparium' of a crop species can be exploited by the plant breeder, is dependent upon the distribution of the genetic discontinuities that are commonly accepted as species differences. Within a species, breeding is straightforward, and it is here that the plant breeder makes the greater part of his gains. Between species, and where sterility barriers have arisen, not only is hybridisation more difficult, but also the genetic understanding and technique required is greater if new and superior types are to be successfully generated.

Breeding in a changing situation
The changing demands of agriculture to which the plant breeder must respond are well illustrated by the history of crop improvement in India. The rate of change at

present, in the era of the 'Green Revolution', is greater than ever before. It is necessary to appreciate that the new cereal varieties are primarily a response to, rather than a cause of, these revolutionary changes. The advent of the new wheats and the new rices was preceded by fifteen years of rapid advance in the productivity of Indian agriculture (Hutchinson, 1971a). The basis of the change was a revolution in fertility on irrigated lands, and the change in cereal varieties took place in response to the opportunities offered by enhanced fertility levels. The plant breeder breeds to exploit the environment available, and in India the breakthrough on the irrigated wheat lands of the northwestern states will spread over the country as fertility levels and water management are improved, and a demand for new varieties is thereby created.

The challenge of the New Agriculture

J.B. HUTCHINSON
St John's College, Cambridge

The evolutionary changes in Indian crop plants that are here recorded, have gone
on in response to the needs of an agricultural system that changed only slowly over
the past 4500 years. The rate of change has increased enormously in the last twenty
years, and the needs of the new agriculture for suitable varieties will only be met
in so far as genetic change matches changes in the crop environment. The adoption
by Indian agricultural scientists of new varieties of wheat from Mexico and rice
from the Philippines was the first, and the most striking, example of a planned
response to the demands of a new farming environment. If this initiative is to be
followed up and extended, some analysis is necessary of the pattern of change in
which Indian agriculture is involved.

The function of plant breeding is to produce varieties that exploit fully the
potential of the environment. In this sense, the breeder's objective is to produce a
variety with a high yield potential, not one with a high yield. Kohli's (1969) data
on the interaction of variety and fertility level in wheat crops illustrate this principle.
At the low fertility levels characteristic of Indian agriculture before the fertiliser
revolution, the old Indian varieties outyielded the new Mexican dwarfs. The
Mexican wheats, however, had a potential that the old varieties had not, and at high
fertiliser levels they greatly outyielded the Indian varieties.

The crop variety is, in fact, only one of four major factors in crop production, all
of which must be advanced together if agricultural productivity is to be increased.
These are: soil fertility, water supply, crop variety and the social factors affecting
the farmer. The success of the wheat revolution in northern India came from the
circumstance that all these were, in fact, in a state of change. Fertility was raised
with fertilisers. Water supply was improved by the exploitation of ground water
supplies with tube wells. Fertility-responsive dwarf wheats became available. Not
least important, an enterprising race of farmers took advantage of the advisory aids
and the credit available to them, and put together fertilisers, water supplies, and
new seed to generate the new level of productivity.

The so-called Green Revolution is fundamentally a fertility revolution (Hutchinson, 1971a). The part of the plant breeder in it has been to supply new varieties
capable of exploiting the higher fertility levels that can now be established. The
evolutionary changes that he is now called upon to effect are all concerned with the
exploitation of the new crop environments that will result from the progressive
improvement of fertility levels. The comparatively unimportant contribution of the
new rice varieties to the increase in rice production – which doubled between 1950
and 1970 – is due to the complexity and diversity of the water supply factors in

rice production. Under the circumstances of the rice crop, no narrow range of new varieties can sweep through the crop as a few Mexican dwarf wheats swept through the irrigated wheat crop. The potential of fertility-responsive rices will only be realised as the small group of introduced types is diversified by breeding within India to meet diverse Indian demands. Indeed it is important to bear this in mind with the wheats also, in that only one half of the wheat acreage is irrigated (Rao, p. 33), and the Mexican wheats have had little impact on the rain-fed areas.

In like manner, the initiation of a Green Revolution on the rain-fed farm lands of India will depend on the application of principles governing high production agriculture to a great range of practical farming circumstances. The principles involved are essentially: good varietal response to fertility improvement; genotypic adaptation to water supply expectation; and storage, nutritional and cooking characteristics to meet consumer requirements.

In the particular circumstances of farming on monsoon rains, the paramount need is for varieties that match in crop duration the period during which there is good expectation of reasonable moisture supplies. This, in general, means early varieties, and with earliness must be associated short stature and high crop index to give the potential to respond to high fertility. In quality characteristics, consumer preference is likely to be met by breeding to the standards of current local varieties; but in nutritional factors it is now possible to breed for high protein content and better amino acid balance than is to be found in current stocks. Further, in some special cases where toxic factors are a problem it is now possible to foresee a substantial amelioration of food hazards, some of which are known to have persisted for many centuries.

In sum, we require to plan for the breeding of a new range of crop plant varieties, not just to create a more productive agriculture, but to exploit the more productive farming systems that are now being devised. This is a major genetic enterprise. The opportunity to breed a new stock that will be well adapted over half a continent only occurs rarely, and when it does it must be followed by diversification to provide varieties that will exploit local potential. Hutchinson and Panse's (1936) conclusion is still valid: 'that in most areas and with most crops, local adaptation is such that local breeding is necessary if the possibilities of crop improvement by improvement of strain are to be fully exploited'. The new circumstance, of which they could not then be aware, is that farming potential is now rapidly improving, and locally adapted material must be regarded as but one of the stocks on which to breed the varieties by which that potential will be exploited.

References

Agogino, G. (1957) *Sci. News Lett., Washington,* 72: 345.

Agogino, G. and Feinhandler, S. (1957) *Texas J. Sci.* 9: 154.

Agogino, G. and Hester, J. (1958) *Amer. Antiquity,* 24: 187.

Agrawal, D.P. (1964) *Science,* 143: 950.

Agrawal, D.P. (1969) *Proc. Archeol. Soc. Ind. Patna.*

Agrawal, D.P. (1971) C-14 Date List, March 71. Mimeographed Circular, Tata Institute of
   Fundamental Research Bombay-5.

Agrawal, D.P., Gupta, S.K. and Kusumgar, S. (1964) *Curr. Sci.* 33: 266.

Agrawal, D.P., Gupta, S.K. and Kusumgar, S. (1969) *Radiocarbon,* 11: 502.

Agrawal, D.P. and Kusumgar, S. (1966) *Radiocarbon,* 9: 451.

Agrawal, D.P. and Kusumgar, S. (1968a) *Curr. Sci.* 37: 96.

Agrawal, D.P. and Kusumgar, S. (1968b) *Radiocarbon,* 10: 134.

Agrawal, D.P. and Kusumgar, S. (1969a) *Radiocarbon,* 11: 190.

Agrawal, D.P. and Kusumgar, S. (1969b) *Curr. Sci.* 38: 5.

Agrawal, D.P., Kusumgar, S., Lal, D. and Sarna, R.P. (1964) *Radiocarbon,* 6: 231.

Aiyer, A.K.Y.N. (1956) *The antiquity of some field and forest flora of India.* The Bangalore
   Press, Bangalore.

Allchin, F.R. (1968) In: Ucko P.J. and Dimbleby, G.W. (eds.), *The domestication and
   exploitation of plants and animals.* Duckworth, London.

Anderson, E. (1943) *Ann. Mo. Bot. Gard.* 30: 469.

Anderson, E. (1945) *Chron. Bot.* 9: 88.

Anderson, E. (1949) *Introgressive hybridisation.* John Wiley and Sons, New York.

Anderson, E. and Sax, K. (1936) *Bot. Gaz.* 97: 433.

Anon. (1963) Report on the marketing of wheat in India. *Agric. Marketing Series* no. 143.

Anon. (1968) Five years of research on dwarf wheats. *I.A.R.I. Bull.* Printed by INSDOC
   Delhi-12.

Bailey, L.H. (1949) *Manual of cultivated plants.* Macmillian & Co., New York.

Bains, S.S., Singh, K.N., Dayanand and Bakshi, J.S. (1970) *Ind. J. Agron.* 15: 356.

Bakshi, J.S. and Luthra, J.K. (1970) The inheritance of stripe rust resistance in barley
   (*Puccinia striiformis West*). 2nd *Int. Barley Genet. Symp.* Pullman, Washington.

Banerjee, S.K. and Swaminathan M.S. (1964) *Ind. J. Genet. & Pl. Br.* 24: 252.

Barber, C.A. (1915) *Mem. Dept. Agr. Ind. Bot. Ser.* 7: 1.

Barber, C.A. (1917) *Mem. Dept. Agr. Ind. Bot. Ser.* 9: 133.

Beasley, J.O. (1940) *J. Hered.* 31: 39.

Beasley, J.O. (1942) *Genetics,* 27: 25.

Bhaduri, P.N. (1933) *J. Ind. Bot. Soc.* 12: 56.

Bhaduri, P.N. and Ghosh, P.N. (1954a) *Nature, Lond.* 174: 934.

Bhaduri, P.N. and Ghose, P.N. (1954b) *Stain Technol.* 29: 269.

Bhaduri, P.N. and Natarajan, A.T. (1956) *Ind. J. Genet. & Pl. Br.* 16: 77, 85.

Bharadwaj, B.D. (1960) Study of hybrid vigour in varietal crosses of maize. Unpublished
    M.Sc. thesis, I.A.R.I., New Delhi.
Bhat, B.K. and Dhawan, N.L. (1969) *Ind. J. Genet. & Pl. Br.* **29**: 321.
Bor, N.L. (1960) *The Grasses of Burma, Ceylon, India and Pakistan.* London.
Bose, R.D. (1931) *Ind. J. Agri. Sci.* **1**: 58.
Bremmer, G. (1966) *Genetica,* **37**: 345.
Buck, E.J. (1925) *Simla Past and Present,* 2nd ed. The Times Press, Bombay.
Burkill, I.H. (1935) *A dictionary of the economic products of the Malay Peninsula.* vol. 1,
    p. 629. Gov. Malaysia and Singapore.
Burkill, I.H. (1953) *Proc. Linn. Soc., London,* **164**: 12.
Burt, B.C. (1913) *Agric. J. India,* **8**: 339.
Buth, G.M. and Chowdhury, K.A. (1971) *Proc. Silver Jubilee Palaeobot. Conf., Lucknow:* 13.

Chanda, S. and Mukherjee, B.B. (1969) *Sci. & Cult.* **35**: 275.
Chandrasekharan, S.N. *et al.* (1946) *J. Ind. Bot. Soc.* **25**: 103.
Chennaveeraiah, M.S. and Patil, S.R. (1968) *Genet. Iberica,* **20**: 23.
Chopra, R.N., Nayar, S.L. and Chopra, I.C. (1956) *Glossary of Indian medicinal plants.* New
    Delhi.
Chowdhury, K.A. and Buth, G.M. (1971) *Biol. J. Linn. Soc.* **3**: 303.
Chowdhury, K.A. and Ghosh, S.S. (1954–5) *Ancient India,* **10, 11**: 121.
Chowdhury, K.A. and Ghosh, S.S. (1955) *Trans. Bose Res. Inst.* **20**: 80.
Chowdhury, K.A., Saraswat, K.S., Hasan, S.N. and Gaur, R.C. (1971) *Sci. & Cult.* **37**: 531.
Clark, A.J. (1936) Improvement in Wheat. *Year Book of Agric.* U.S.D.A.: 207.
Classen, C.E. and Hoffman, A. (1950) *J. Amer. Soc. Agron.* **42**: 79.
Clayton, W.D. (1968) *Kew Bull.* **21** (3): 485.
Collins, G.N. (1909) *U.S.D.A. Bur. Plant Industry Bull.* **16**: 1.
Covas, G. (1950) *Rev. Argent. Agron.* **17**: 257.
C.S.I.R.I. (Council of Scientific and Industrial Research, India) (1948) *Wealth of India.*

Darlington, C.D. (1963) *Chromosome botany and the origins of cultivated plants.* George Allen
    & Unwin, London.
Darlington, C.D. and Wylie, A.P. (1956) *Chromosome atlas of flowering plants.* George Allen
    & Unwin, London.
Datta, P.C. and Naug, A. (1968) *Beitraege zur Biologie der Pflanzen,* **45**: 113.
de Candolle, A. (1886) *Origin of cultivated plants.*
Deodikar, G.B. and Thakar, C.V. (1956) *Proc. Ind. Acad. Sci. (B),* **43**: 37.
Desai, B.B. (1927) *Agric. J. India,* **22**: 351.
Dhawan, N.L. (1964) *Maize Genet. Coop. News Letter,* **38**: 69.
Doggett, H. (1970) *Sorghum.* Longman, London and Harlow.
Douwes, H. (1951) *J. Genet.* **50**: 179.

Ecklow, C.F. and Zeyher, C. (1836) *Enumeratio plantarum Africae Australies extra-tropicae.*
Ellerton, S. (1939) *J. Genet.* **38**: 307.
Ellison, W. (1936) *J. Genet.* **32**: 473.
Engler, A. (1915) *Pflanzenwelt Afrika* 3(1): 665.
Engler, A. and Prantl, K. (1889) *Die natürlichen Pflanzen familien,* II Teil.
Erdtman, G. (1956) *Grana Palynologica,* **1**: 127.
Erdtman, G, Praglowski, J. and Radwan (1959) *Bot. Notiser.* **112**: 176

Faegri, K. and Iversen J. (1964) *A. Textbook of Pollen Analysis.*
Fluckiger and Hanbury (1892) In: Pharmacography 567, *Dictionary of Economic Products of
    India,* vol. 6 part 1.

Ford, C.E. (1938) *Genetica,* **20**: 431.

Gadwal, V.R. (1966) Studies in interspecific hybridisation in the genus *Abelmoschus.* Unpublished thesis, I.A.R.I., New Delhi.

Gadwal, V.R., Joshi, A.B. and Iyer R.D. (1968) *Ind. J. Genet. & Pl. Br.* **28**: 269.

Ghosh, S.S. (1961) *Ind. Forest,* 87: 295.

Ghosh, S.S. and Lal, Krishna (1962–3) *Ancient India,* **18**: 161.

Gode, P.K. (1945) *Annals (B.O.R. Inst.),* 27: 89.

Gode, P.K. (1946) *Pracyavani,* Jan.–Apr.: 35.

Govindaswamy, S., Krishnamurty, A. and Sastry, N.S. (1966) *Oryza,* **3**: 74.

Grant, W.F. (1959) *Canad. J. Genet. Cytol.* **1**: 313.

Grohne, U. (1957) *Photogr. Und. Forsch.* 7.

Gulati, A.N. (1961) *Technical reports on archaeological remains.* Poona.

Gulati, A.N. (1965) In: Deo, S.B. and Ansari, D.D. (eds.), *Chalcolithic Chandoli.*

Günther, E. (1959) Cytologische untersuchungen an *Solanum nigrum* L. *Ber. dt. bot. Ges.* **72**: 14.

Gupta, D. and Jain, H.K. (1971*a*) *Maize Genet. Coop. News Letter* **45**: 37.

Gupta, D. and Jain, H.K. (1971*b*) *Maize Genet. Coop. News Letter,* **45**: 39.

Habib, I. (1963) *The agrarian system of Mughal India* (1556–1707) Asia Pub. House, India.

Harberd, D.J. (1972) *Bot. J. Linn. Soc.* **65**: 1.

Hardas, M.W. and Joshi, A.B. (1954) *Ind. J. Genet & Pl. Br.* **14**: 47.

Harlan, J.R. (1968) On the origin of barley. Agriculture Handbook no. 338, U.S. Dept. of Agriculture.

Harlan, J.R. (1969) On the origin of barley. A second look, *2nd Int. Barley Genet. Symp.* Pullman, Washington.

Harland, S.C. (1936) *Biol. Rev.* **11**: 82.

Harland, S.C. (1939) *The Genetics of Cotton.* Jonathan Cape, London.

Helbaek, H. (1970) *The plant husbandry of Hacilar. Excavations at Hacilar.* Edinburgh University Press.

Herklots, G.A.C. (1972) *Vegetables in South-East Asia.* George Allen & Unwin, London.

Hiern, W.P. (1896) *Catalogue of the African plants.*

Hilterbrandt, U.M. (1935) *Lenin Acad. Agric. Sci. Inst. Plant Industry, Moscow and Leningrad,* no. 6, p. 55.

Himada, H. (1956) Land and crops of Nepal, Himalayas. *Japan.*

Hochreutiner, B.P.G. (1900) *Ann. Cons. Jard. Bot. Geneve,* **4**: 23.

Hochreutiner, B.P.G. (1924) *Candollea,* **2**: 79.

Hooker, J.D. (1854) *Himalayan Journals,* vols. I and II. John Murray, London.

Hooker, J.D. (1872) *Flora of British India,* vol. I. Gov. Ind. Publ.

Hooker, J.D. (1884) *Flora of British India,* vol. IV. Gov. Ind. Publ.

Hooker, J.D. (1890) *Flora of British India,* vol. V. Gov. Ind. Publ.

Howard, A. and Howard, G.L.C. (1909*a*) *Mem. Dept. Agric. Ind. Bot. Ser,* **2**: 7.

Howard, A. and Howard, G.L.C. (1909*b*) *Wheat in India: its production, varieties and improvement.* Calcutta.

Howard, A., Howard, G.L.C. and Rehman Khan, A.R. (1910) *Mem. Dept. Agric. Ind. Bot. Ser.* **3**: 281.

Howard, A., Howard, G.L.C. and Rehman Khan, A.R. (1922) *Mem. Dept. Agric. Ind. Bot. Ser.* **12**: 20.

Howard, G.L.C. (1916) *Mem. Dept. Agric. Ind. Bot. Ser.* **8**: 88.

Hutchinson, J. and Dalziel, K. (1927) *Flora of Western Tropical Africa.*

Hutchinson, J.B. (1951) *Heredity,* 5, 161.

Hutchinson, J.B. (1954) *Heredity*, 8: 225.

Hutchinson, J.B. (1959) *The application of genetics to cotton improvement*. Cambridge University Press, London.

Hutchinson, J.B. (1965) In: Hutchinson, J.B. (ed.), *Essays in crop plant evolution*. Cambridge University Press, London.

Hutchinson, J.B. (1970) *Ind. J. Genet. & Pl. Br.* 30: 269.

Hutchinson, J.B. (1971*a*) *J. Roy. Soc. Arts,* 119: 104.

Hutchinson, J.B. (1971*b*) *Expl. Agric.* 7: 273.

Hutchinson, J.B. and Ghose, R.L.M. (1937) *Ind. J. Agric. Sci.* 7: 1.

Hutchinson, J.B. and Panse, V.G. (1936) *Agric. & Livestock in India,* 6: 397.

Hutchinson, J.B., Silow, R.A. and Stephens, S.G. (1947) *The evolution of Gossypium,* Oxford University Press, London.

Iversen, Johs. (1941) *Denm. Geol. Unders.* 66: 1.

Jacob, K.M. (1957) *Cytologia,* 22: 380.

Jain, H.K. and Gupta, D. (1971*a*) *Maize Genet. Coop. News Letter,* 45: 47.

Jain, H.K. and Gupta, D. (1971*b*) *Maize Genet. Coop. News Letter,* 45: 50.

Jain, H.K. and Gupta, D. (1971*c*) *Maize Genet. Coop. News Letter,* 45: 53.

Jeffreys, M.D.W. (1965) *Anthropol. J. Canada,* 3: 5.

Johnson, G.W. (1847) *The Gardener,* 1: 19.

Jørgensen, C.A. (1928) *J. Genet.* 19: 133.

Joshi, A.B. and Hardas, M.W. (1953) *Curr. Sci.* 12: 384.

Joshi, A.B. and Hardas, M.W. (1956) *Nature, Lond.* 178: 1190.

Joshi, J.P. (1972) *Proc. Int. Symp. Radiocarbon & Ind. Archaeology.* Tata Inst. of Fundamental Res., Bombay.

Khoshoo, T.N. (1971) *Ind. J. Genet. & Pl. Br.* 31: 305.

Khoshoo, T.N. and Pal, M. (1972) *Chromosomes Today,* 3: 259.

Kohli, S.P. (1969) Wheat varieties in India. *I.C.A.R. Tech. Bull. (Agric.),* no. 18, New Delhi.

Koul, A.K. and Paliwal, R.L. (1964) *Cytologia,* 29: 375.

Koul, A.K. and Paliwal, R.L. (1965) *Kash. Sci.* 2: 114.

Krauss, F.G. (1927) *J. Hered.* 17: 227.

Kumar, L.S.S., Thombre, M.V. and D'Cruz, R. (1958) *Proc. Ind. Acad. Sci.* (B), 47: 252.

Kurita, M. (1946) *Japan. J. Genet.* 21: 63.

Kuwada, H. (1966) *Japan. J. Breed.* 16: 21.

Leake, H.M. and Ram Prasad (1912) *Bull. no. 31 Agric. Res. Inst., Pusa.*

Lecomte, M.H. (1910) *Flora générale de L'Indo-Chine,* Masson, Paris.

Longley, A.E. (1939) *J. Agr. Res.* 59: 475.

Luthra, J.C. (1936) *Curr. Sci.* 4: 489.

Mackay, E.J.H. (1943) *Chanhu-Daro Excavations. New Haven, Connecticut.*

Magoon, M.L., Ramanujam, S and Cooper, D.C. (1962) *Caryologia,* 15: 151.

Mahta, D.N. and Dave, B.B. (1931) *Mem. Dept. Agr. Ind. Bot. Ser.* 19: 1.

Mangelsdorf, P.C. (1958) *Science,* 128, 1313.

Mangelsdorf, P.C. and Oliver, D.L. (1951) *Bot. Mus. Leafl., Harvard Univ.* 14: 263.

Mangelsdorf, P.C. and Reeves, R.G. (1939) *Texas Agri. Exp. Sta. Bull.* 574: 1.

Mangelsdorf, P.C. and Reeves, R.G. (1959) *Bot. Mus. Leafl., Harvard Univ.* 18: 413.

Manton, I. (1932) *Ann. Bot.* 46: 509.

Markham, C.R. (1862) *Travels in Peru & India.* John Murray, London.

Marshall, Sir J. (1931) *Mohenjo-daro and the Indus Valley civilization* I. London.

Masur, N.G. (1933) *Agric. and Livestock in India,* 3(2): 125.

Mehra, K.L. (1963*a*) *Curr. Sci.* 32: 300.

Mehra, K.L. (1963*b*) *Phyton,* **20**: 189.

Mehra, K.L. (1968) *Advancing Frontiers of Plant Sciences,* **19**: 51.

Merrill, E.D. (1954) The botany of Cook's voyage. *Chronica Botanica.* Waltham, Massachusetts.

Misra, P.S. (1965) Unpublished Ph.D. thesis, I.A.R.I., New Delhi.

Misra, P.S., Pal, M., Mitra, C.R. and Khoshoo, T.N. (1971) *Proc. Ind. Acad. Sci.* (B), **74**: 155.

Misro, B. and Rath, G.C. (1961) *J. Biol. Sci., Bombay,* **4**: 35.

Moon, P. (1947) *Warren Hastings and British India.* Hodder, London.

Mukherjee, B.K., Gupta, N.P., Singh, S.B. and Singh, N.N. (1971) *Euphytica,* **20**: 113.

Murdock, G.P. (1959) *Africa, its people and their culture and history.* McGraw Hill, New York.

Murty, G.S. (1958) *Wheat varieties. Rev. Bull. I.C.A.R., New Delhi.*

Murty, B.R., Arunachalam, V., Doloi, P.C. and Ram, J. (1972) *Heredity,* **28**, 287.

Nagaraja Rao, M.S. (1971) *Protohistoric cultures of the Tungabhadhra Valley.* Dharwar.

Nagaraja Rao, M.S. and Malhotra, K.C. (1965) *Stone Age Hill Dwellers of Tekkalkota.* Poona.

Nakamura, M. (1937) *Cytologia,* **1**: 57.

Narain, A (1951) *Science & Culture,* **16**: 484.

Narain, A. (1952) *Curr. Sci.* **21**: 166.

Narain, A. (1958*a*) Unpublished Ph.D. thesis, University of Agra.

Narain, A. (1958*b*) *Indian Oilseed J.* **2**: 92.

Narain, A. (1959) *Indian Oilseed J.* **3**: 37.

Narain, A. (1960) *J. Hort. Assoc. Japan.* **29**(4): 331.

Narain, A. (1962) *Indian Oilseed J.* **6**: 154.

Narain, A. (1967) *Nature,* **213**: 198.

Narain, A. (1968) *Proc. 12th Int. Congr. Genetics, Tokyo.*

Narain, A. (1969) *Curr. Sci.* **38**(9): 224.

Narain, A. (1971*a*) *Theoretical & Appl. Genet.,* **41**: 203.

Narain, A. (1971*b*) *Proc. of the First All-India Congr. of Cytol. & Genet., Chandigarh.*

Narain, A. and Singh, P. (1968) *J. Hered.* **59**: 287.

Narain, Lala Aditya (1970) *J. Bihar Res. Soc.* **46**: 1.

Nirodi, N. (1955) *Ann. Miss. Bot. Gd.* **42**: 103.

Olsson, G. (1954) *Hereditas,* **40**: 398.

Pal, B.P. (1966) Wheat. *I.C.A.R. Monograph, New Delhi.*

Pal, B.P. and Pushkarnath (1951) Indian Potato Varieties, *I.C.A.R. Bull. no.* 62.

Pal, B.P., Singh, H.B. and Swarup, V. (1952) *Bot. Gaz.* **13**: 455.

Pal, M. and Khoshoo, T.N. (1972) *J. Hered.* **63**: 78.

Parkash, O. (1961) Food and drinks in ancient India. *Munshi Ram Manohar Lal, Delhi.*

Parthasarathy, N. (1947) M.O.P. Iyengar Commem. Vol. *J. Ind. Bot. Soc.* p. 133.

Patel, C.T. (1971) *Cotton Development,* **1**(2): 1.

Patil, V.P. and Deodikar, G.B. (1968) *Ind. J. Genet. & Pl. Br.* **28**: 342.

Percival, J. (1921) *The wheat plant — A monograph.* Duckworth, London.

Phillips, L.L. (1963) *Evolution,* **17**: 460.

Phillips, L.L. (1966) *Amer. J. Bot.* **53**: 328.

Phillips, L.L. and Strickland, M.A. (1966) *Can. J. Genet. Cytol.* **8**: 91.

Pieris, H.A. (1936) *Trop. Agriculturist, Ceylon,* **86**: 217.

Portères, R. (1956) *J. Agri. Trop. Bot. Appl.* **3**: 341, 541, 627, 821.

Prain, D. (1898) *Agr. Ledger,* **5**: 1.

Prain, D and Burkill, I.S. (1938) *Ann. Roy. Bot. Garden, Calcutta,* **14**, 348.

Price, S. (1968) *Econ. Bot.* **22**: 155.

Purohit, A.N. (1970) *New Phytol.* **69**: 521.

Purseglove, J.W. (1968) *Tropical crops:* Dicotyledons vol. II. Longman, London.

Raikes, R.L. and Dyson, R.H. Jnr. (1961) *Amer. Anthrop.* **63**: 165.

Ramanatha Ayyar, V. (1937) *Proc. 1st Conf. Sci. Workers on Cotton in India.* I.C.C.C., Bombay.

Raychowdhury, S.P. (1964) *Agriculture in ancient India.* I.C.A.R., New Delhi.

Return: Cotton (India) (1847) *Ordered by the House of Commons to be printed.*

Richharia, R.H. (1937) *Ind. Agric. Sci.* **7**: 707.

Ridgway, R. (1912) *Colour standards and colour nomenclature.*

Riley, R. (1965) In: Hutchinson, J.B. (ed.), *Essays on crop plant evolution,* Cambridge.
University Press, London.

Riley R. and Kimber, G. (1966) *Rep. Pl. Br. Inst. 1964—5.*

Robbelen, G. (1960) *Chromosoma,* **11**: 205.

Rowley, John R. (1960) *Grana Palynologica,* **2**: 9.

Roy, B. (1933) *Ind. J. Agr. Sci.* **3**: 1098.

Roy, J.K. (1963) Unpublished Ph.D., thesis, Utkal University, Bhubaneswar.

Roy, R.P. and Jha, R.P. (1958) *Cytologia,* **23**: 356.

Sahni, B. (1936) *Curr. Sci.* **4**: 796.

Sahni, B. (1938) Presidential address. *Proc. 25th Ind. Sci. Congr. Assoc., Calcutta.*

Salaman, R.N. (1946) *J. Linn. Soc. (Bot.),* **53**: 1.

Salaman, R.N. (1954) *J. Linn. Soc. (Bot.),* **55**: 185.

Salaman, R.N. and Hawkes, J.G. (1949) *Proc. Linn. Soc.* **161**: 71.

Sampath, S. (1962) *Oryza,* **1**: 1.

Sampson, H. (1911) *Agric. J. India,* **6**: 365.

Sankalia, H.D., Deo, S.B. and Ansari, Z.D. (1960) *From History to Prehistory at Nevasa (1954—
1956),* Poona.

Sankalia, H.D., Deo, S.B. and Ansari, Z.D. (1969) *The Excavations at Ahar (Tambavati).* Poona.

Sankalia, H.D., Subbarao, B. and Deo, S.B. (1953) *SW. Jour. of Anthrop.* **9**: 343.

Sankalia, H.D., Subbarao, B. and Deo, S.B. (1958) *The Excavations at Navadatoli and
Maheshwar,* Poona.

Sauer, J.D. (1950) *Ann. Missouri Bot. Gard.* **37**: 561.

Sauer, J.D. (1967) *Ann. Missouri Bot. Gard.* **54**: 103.

Schaaffhausen, R.V. (1952) *Econ. Bot.* **6**: 216.

Schumann, K. (1895) In: Engler, A. and Prantl, H., *Die natürlichen Pflanzen familien,*
vol. 3(6), p. 47. Leipzig.

Seetharam, A. and Srinivasachar, D. (1970) *Curr. Sci.* **39**: 492.

Sen, S.N. (1963) Transmission of scientific ideas between India and foreign countries in ancient
and medieval times. Symp. on *The History of Sciences in India.* National Inst. of
Sciences of India, New Delhi.

Sethi, R.L. (1931) *Agriculture and Livestock in India,* **1**(3): 243.

Sharma, S.D. and Shastry, S.V.S. (1971) *Riso.*

Shaw, F.J., Rehman, K. and Singh, H. (1933) *Ind. J. Agr. Sci.* **3**: 1.

Sikdar, A.K. and De, D.N. (1967) *Bull. Bot. Soc. Beng.* **21**: 25.

Sikka, S.M. (1940) *J. Genet.* **40**: 441.

Simmonds, N.W. (1959) *Bananas.* Longman, London.

Simmonds, N.W. (1962) *The Evolution of the Bananas.* Longman, London.

Simmonds, N.W. (1963) *J. Linn. Soc. (Bot.),* **58**: 461.

Simmonds, N.W. (1964) *J. Linn. Soc. (Bot.),* **59**: 43.

Simmonds, N.W. (1968) *Euphytica,* **17**: 504.

Singh, D. (1958) *Rape & Mustard.* I.C.O.C. Hyderabad.

Singh, G. (1967) *Ind. Hydrol.* **3**: 111.

Singh, G. (1970) History of post-glacial vegetation and climate of the Rajasthan desert. Report
submitted to University of Wisconsin, Madison, U.S.A.

Singh, G. (1971) *Archaeol. and Phys. Anthropol. in Oceania,* **6**: 177.

Singh, H. (1961) I.C.A.R., *Cereal Crops Ser.* 1: 1.

Singh, H.B. and Dadlani, S.A. (1967) *Ind. Fmg.* 17: 4.

Singh, Ram Dhan (1946) *Ind. J. Genet. & Pl. Br.* 6: 34.

Singh, S.R. and Narain, A. (1965) *Ind. Oilseed J.* 9: 215.

Sinha, S.K. and Pushkarnath (1964) *Ind. Potato J.* 6: 24.

Skovsted, A. (1934) *J. Genet.* **23**: 407.

Skovsted, A. (1935) *J. Genet.* **31**: 263.

Skovsted, A. (1937) *J. Genet.* **34**: 97.

Snowden, J.D. (1936) *The cultivated races of Sorghum.* Adlard.

Stebbins, G.L. (1947) *Ecol. Monog.* 17: 149.

Stebbins, G.L. (1950) *Variation and evolution in plants.* Columbia University Press, New York.

Stebbins, G.L. and Paddock, E.F. (1949) *Madrono,* **10**: 70.

Stevenson, G.L. (1965) *Genetics and breeding of sugarcane.* Longman, London.

Stonor, C.R. and Anderson, E. (1949) *Ann. Mo. Bot. Gard.* **36**: 355.

Subramaniam, N. and Srinivasan, M. (1952) *Proc. Soc. Biol. Chemists, India,* **10**: 25.

Swaminathan, M.S. (1949) Cytotaxonomic studies in the Genus *Solanum.* Unpublished thesis, I.A.R.I., New Delhi.

Swaminathan, M.S. (1958) *Ind. J. Genet. & Pl. Br.* **18**: 8.

Swaminathan, M.S. (1963) *Proc. 2nd Int. Wheat Genet. Symp. Hereditas Lund. suppl.*

Swaminathan, M.S., Magoon, M.L., and Mehra, K.L. (1954) *Ind. J. Genet & Pl. Br.* **14**: 87.

Tandon, S.L. and Rao, G.R. (1966*a*) *Ind. J. Genet. & Pl. Br.* **26**: 130.

Tandon, S.L. and Rao, G.R. (1966*b*) *J. Cytol. Genet.* 1: 41.

Terry, E. (1655) *Voyage to East India.* J. Wilkie, London, reprinted in 1777 from the edition of 1655.

Teshima, T. (1933) *J. Fac. Agric. Hokkaido Univ.* **34**: 1.

Thapa, J.K. (1966) *Bull. Tech.* 3: 29. Namgyal Inst. of Technology, Gangtok, Sikkim.

Thapar, B.K. (1972) *Proc. Int. Symp. Radiocarbon & Ind. Archaeology. Tata Inst. of Fundamental Res. Bombay.*

Thoday, J.M. (1964) *Genetics Today,* **3**: 533. Pergamon Press, London & Oxford.

Thomas, G.P. (1846) *Views of Simla.* Dickinson & Co., London.

Thonner, Fr. (1915) *The Flowering Plants of Africa.* Dulau & Co., London.

Tschechow, W. and Karataschowa, N. (1932) *Cytologia,* **3**: 221.

Tucker, J.M. and Sauer, J.D. (1958) *Madrono,* **14**: 252.

Upadhya, M.D., Purohit, A.N. and Sharda, R.T. (1972) *World Crops,* **24**: 314.

Vallaeys, G. (1948) *Bull. Agric. Congo Belge,* **39**: 247.

Vats, M.S. (1940) *Excavations at Harappa I.* Calcutta.

Vaughan, J.H., Hemingway, J.S. and Schofield, H.J. (1965) *J. Linn. Soc. (Bot.),* **58**: 374.

Vavilov, N.I. (1926) *Trudi poPrikil Bot.* 17(2).

Vavilov, N.I. (1951) *Chron. Bot.* 13: 1.

Vishnu-Mittre (1962) *Tech. Report on Archaeol. Remains,* p. 11. Deccan College, Poona.

Vishnu-Mittre (1966*a*) *Palaeobot.* 15: 142.

Vishnu-Mittre (1966*b*) *Palaeobot.* 15: 157.

Vishnu-Mittre (1967—8) *Puratatva,* 1: 4.

Vishnu-Mittre (1968*a*) In: Khazanchi, T.N., *Report on Excavations at Burzahom.*

Vishnu-Mittre (1968*b*) *History of Agriculture in India.* Deccan College and Post-graduate Research Institute, Poona.

Vishnu-Mittre (1968*c*) Protohistoric Records of Agriculture in India. *J.C. Bose Endowment Lecture. Trans. Bose Res. Inst. Calcutta,* 31.

Vishnu-Mittre (1968d) In: Dikshit, M.G. (ed.) *Excavations at Kaundinyapura.* Bombay.

Vishnu-Mittre (1969) In: Sankalia, H.D. *et al.* (ed.), *Excavations at Ahar (Tambavati).* Poona.

Vishnu-Mittre (1970) *Ind. J. History Sci.* **5**: 143.

Vishnu-Mittre (1971a) In: Nagaraja Rao, M.S. (ed.), *Protohistoric Cultures of the Tungabhadra Valley.* Dharwar, India.

Vishnu-Mittre (1971b) *Proc. 3rd Int. Palynol. Conf., Novisibirsk, U.S.S.R.*

Vishnu-Mittre (1971c) *Proc. Silver Jubilee Palaeobot. Conf., Lucknow,* 72.

Vishnu-Mittre (1972) *Palaeobot.* **21.**

Vishnu-Mittre and Gupta, H.P. (1966) *Palaeobot.* **15**: 176.

Vishnu-Mittre and Gupta, H.P. (1968) In: Deo, S.B. and Dhavalkar, M.K. (eds.), *Paunar excavations – 1967.* Nagpur.

Vishnu-Mittre and Gupta, H.P. (1968–9) *Puratatva,* **2**: 21.

Vishnu-Mittre and Gupta, H.P. (1971) *Palaeobot.* **19**: 110.

Vishnu-Mittre, Gupta, H.P. and Robert, R.D. (1967) *Curr. Sci.* **36**: 539.

Vishnu-Mittre, Prakash, U. and Awasthi, N. (1972) *Geophytol.* **1**: 170.

Vishnu-Mittre and Sharma, B.D. (1966) *Palaeobot.* **15**: 185.

Waalkes, J.v.B. (1966) *Blumea,* **14**: 1.

Watt, G. (1889) *A dictionary of the economic products of India,* vols. 1 and 2.

Watt, G. (1892) *A dictionary of the economic products of India,* Vol. 6.

Watt, G. (1904) *Agr. Ledg.* **11**: 189.

Watt, G. (1907) *The wild & cultivated cotton plants of the World.* Longman, London.

Watt, G. (1908) *The commercial products of India.* John Murray, London.

Weatherwax, P. (1954) *Indian corn in Old America.* Macmillan, London.

Webber, J.M. (1939) *J. Agric. Res.* **58**: 237.

West, R.G. (1956) *Phil. Trans. Roy. Soc.* B, **239**: 265.

Westergaard, M. (1948) *K. Danske Videnskabernes Selskab,* **18**: 1.

White, G.F. (1838) In: Roberts, Emma (ed.), *Views in India, chiefly among the Himalaya Mountains.* Fisher, Son & Co., London and Paris.

White, O.P. (1918) *J. Hered.* **9**: 195.

Zimmerman, L.H. and Parkey, W. (1954) *Agron J.* **46**, 287: c.f. *Advances in Agronomy,* **10.** (1958).

Zukovsky, P.M. (1962) *Cultivated plants and their wild relatives.* Commonwealth Agricultural Bureaux, London.

Index

174